Harmonisation of Regulatory Oversight in Biotechnology

Safety Assessment of Transgenic Organisms in the Environment, Volume 10

OECD CONSENSUS DOCUMENT ON ENVIRONMENTAL CONSIDERATIONS FOR THE RELEASE OF TRANSGENIC PLANTS

OECD
BETTER POLICIES FOR BETTER LIVES

Please cite this publication as:
OECD (2023), *Safety Assessment of Transgenic Organisms in the Environment, Volume 10: OECD Consensus Document on Environmental Considerations for the Release of Transgenic Plants*, Harmonisation of Regulatory Oversight in Biotechnology, OECD Publishing, Paris, https://doi.org/10.1787/62ed0e04-en.

ISBN 978-92-64-89796-0 (print)
ISBN 978-92-64-62570-9 (pdf)
ISBN 978-92-64-18271-4 (HTML)
ISBN 978-92-64-35168-4 (epub)

Harmonisation of Regulatory Oversight in Biotechnology
ISSN 2414-6854 (print)
ISSN 2311-4622 (online)

Photo credits: Cover © Kondrashin Sergey/Shutterstock.com.

Corrigenda to OECD publications may be found on line at: www.oecd.org/about/publishing/corrigenda.htm.

Foreword

From their first commercialisation in the mid-1990s, genetically engineered crops (also known as "transgenic" or "genetically modified" plants) have been approved for commercial release in an increasing number of countries, for planting, entering in the composition of foods and feeds, or use in industrial processing. Most of these productions to date are for soybean, maize, cotton and rapeseed (canola) bearing pest resistance and herbicide tolerance traits, aiming to improve yields and reduce the costs of production. Other transgenic crops are increasingly grown, and other traits increasingly introduced in engineered plants, adapting them to biotic or abiotic stress, such as resistance to drought or tolerance to salt in the growing environment, or changing a characteristic, e.g., modified oil content, reduced lignin content, non-browning or nutritional quality (biofortification). Thus, transgenic plants, where adopted and available on the market, enlarge possibilities for farmers, industry and consumers. They can play a part in addressing global concerns such as the rising need for food and feed in the growing population context, or the necessary adaptation of agriculture for better resilience to climate change.

Modern biotechnologies are applied to plants (crops, flowers, trees), animals and micro-organisms. The safety of the resulting genetically engineered organisms, when released in the environment for their use in agriculture, forestry, fishery, the food and feed industry, biofuel production or other applications, represents a challenging issue. A scientifically sound approach to their risk assessment should inform biosafety regulators and support national decisions regarding their possible market release. Genetically engineered products are rigorously assessed by their developers during their elaboration and by governments when ready for commercial use, to ensure high safety standards for the environment, human food and animal feed. Such assessments are considered essential for healthy and sustainable agriculture, industry and trade.

The OECD offers long-standing recognised expertise in biosafety and contributes to facilitating a harmonised approach. Since 1995, the OECD Working Party on the Harmonisation of Regulatory Oversight in Biotechnology (WP-HROB) has brought together national authorities responsible for the environmental risk/safety assessment of transgenic products in OECD countries and partners. Other international organisations involved in biosafety activities are associated with this programme.

The primary goals of the WP-HROB are to promote international regulatory harmonisation and ensure that methods used in the risk/safety assessment of genetically engineered products are as similar as possible. This opens the way to recognition and possible acceptance of information from the assessments of other countries. The benefits of harmonisation are multiple: it strengthens mutual understanding among countries, prevents duplication of efforts, saves resources and increases the efficiency of the risk assessment process. Overall, it improves safety while reducing unnecessary barriers to trade.

Guidance and tools developed by the WP-HROB to help the environmental risk/safety assessment of transgenic organisms (or "biosafety") are already being used worldwide. Biosafety consensus documents are major outputs of its work, addressing the key elements and core set of science-based issues that countries believe are relevant to biosafety assessments. Being publicly available, these documents can also benefit other countries around the world wishing to use these tools following the same principles.

In addition, information on the transgenic plants approved for commercial release in at least one country for use in agriculture and/or in foods and feeds processing can be found in the OECD BioTrack Product Database (https://biotrackproductdatabase.oecd.org). Each transgenic product is described, with information on approvals in different countries. To date, this database covers 393 varieties from 26 plant species approved in 18 countries/regions and continues to expand.

The fast development and increasing use of a range of new breeding techniques, including "genome editing", allows for quicker and more efficient development of applications at a lower cost. These techniques are being reviewed by regulators, risk assessors, researchers and developers for their potential impact on risk/safety assessment while favouring a coherent policy approach to facilitate innovation, and the OECD offers the relevant platform for it (e.g., see the proceedings of the OECD conference "Genome Editing: Applications in Agriculture – Implications for Health, Environment and Regulation" held in 2018).

More than sixty consensus and guidance documents have been published by the WP-HROB to date. Their scope is growing in line with the new biotechnological developments and wider applications to new fields. The list shown in Annex H of the publication summarises the extent of the species or subjects currently covered and in which volume of the series to find them. In the area of plants, these science-based publications deal with the biology of crop and tree species, selected traits introduced into plant species, and other biosafety issues arising from modifications made to plants. They are available at www.oecd.org/biotrack.

This Volume 10, containing the OECD Consensus Document on Environmental Considerations for Risk/safety Assessment for the Release of Transgenic Plants, is of different content, dealing with environmental risk/safety assessment at a broader level. The document contains general information on points that risk/safety assessors should focus on when planning assessments for the release of transgenic plants into the environment. The annexes describe seven examples of environmental considerations routinely examined by assessors and taken from actual experiences gained during risk/safety assessment of transgenic plants.

The purpose of this document is not to elaborate new terminology or to describe how to undertake an actual risk/safety assessment, but rather to describe an approach and provide illustrative examples for planning and structuring an environmental risk/safety assessment. The set of science-based information contained in this Volume, previously agreed by consensus and published by the OECD, constitute a solid reference recognised internationally, and a tool for use during the environmental risk and safety assessment process. The document should be of interest to regulators and assessors in charge of evaluating the risk/safety of transgenic plants prior to environmental release, as well as to plant breeders and the wider scientific community.

Complementing the biosafety work developed at OECD, the programme on the safety of novel foods and feeds develops guidance and consensus documents on the composition of foods and feeds derived from transgenic plants that can be used in a comparative approach. More information on the novel food and feed safety programme can be found on the OECD BioTrack website (www.oecd.org/biotrack).

The consensus documents published in Volumes 1 to 10 of the Series are also available individually free of charge on the OECD BioTrack website. The WP-HROB endorsed this document, which is published under the responsibility of the Chemical and Biotechnology Committee of the OECD.

Acknowledgments

This publication results from the common effort of the participants of the OECD Working Party on the Harmonisation of Regulatory Oversight in Biotechnology (WP-HROB), concluding a long-standing project achieved in 2023. It contains the OECD Consensus Document on Environmental Considerations for Risk/safety Assessment for the Release of Transgenic Plants, a document of major importance dealing with the 'core' of the biosafety work, justifying its publication as a stand-alone Volume 10 of the series of compendium publications.

Canada served as the lead of the Steering Group that drove the development of this document, followed by the Bureau of the WP-HROB from 2020 onward. The various parts of the document were each drafted by a sub-group of relevant experts and authorities, under the leadership of Canada (main body); Australia (Annex A); Mexico and then the United States (Annex B); Belgium (Annex C); the Netherlands (Annex D); United States (Annex E); Canada (Annex F); and Finland (Annex G). The draft document also benefitted from successive inputs from other WP-HROB delegates and experts, whether from OECD member countries, non-member economies and observer organisations.

Philip Mcdonald (Canada) was instrumental in the original idea for this document and that it could be framed around a set of common 'environmental considerations'. The contribution of the late Dr Alan Raybould (BIAC) regarding the use of the 'problem formulation approach' is also acknowledged.

This volume 10 was prepared by Akihiro Kagoshima under the supervision of Bertrand Dagallier, at the EHS Division of the OECD Environment Directorate. The OECD is grateful to the scientists, regulators and authorities who participated in the development of the body text and the seven Environmental Considerations, and wishes to thank each of them.

Table of contents

Tables

Boxes

Abbreviations and acronyms

AFSI	Agriculture and Food Systems Institute
AUDA-NEPAD	African Union Development Agency – New Partnership for Africa's Development
BIAC	Business at OECD
DNA	Deoxyribonucleic acid
dsRNA	Double-strand Ribonucleic acid
EFSA	European Food Safety Authority
FAO	Food and Agriculture Organization of the United Nations
ILSI-CERA	International Life Science Institute – Center for Environmental Risk Assessment
ILSI-RF	International Life Science Institute – Research Foundation
IPCS	International Programme on Chemical Safety
IPPC	International Plant Protection Convention
OECD	Organisation for Economic Co-operation and Development
Rdna	Recombinant Deoxyribonucleic acid
SCBD	Secretariat of the Convention on Biological Diversity
UNEP	The United Nations Environment Programme
WP-HROB	OECD Working Party on the Harmonisation of Regulatory Oversight in Biotechnology
WP-SNFF	OECD Working Party for the Safety of Novel Foods and Feeds

Executive summary

This document constitutes the tenth volume of the OECD Series on Harmonisation of Regulatory Oversight in Biotechnology, which relates to the environmental risk/safety assessment of transgenic organisms, also called "biosafety". The Series collate individual "consensus documents" published by the Working Party on the Harmonisation of Regulatory Oversight in Biotechnology. The nine previous volumes covered documents issued from 1996 to 2022. The current volume contains the Consensus Document on Environmental Considerations for Risk/safety Assessment for the Release of Transgenic Plants, published in 2023.

Modern biotechnologies are applied to plants (crops, flowers, trees), animals and micro-organisms. The safety of the resulting transgenic organisms, when released in the environment for use in agriculture, forestry, fishery, the food and feed industry, biofuel production or other applications, represents a challenging issue. Genetically engineered products are rigorously assessed by their developers and by governments to ensure high safety standards. These risk/safety assessments, conducted through a scientifically sound approach, inform biosafety regulators and support the decision concerning the release of novel organisms in the environment.

The OECD offers long-standing recognised expertise in biosafety and contributes to facilitating a harmonised approach. The OECD consensus documents identify information of relevance to the environmental risk/safety assessment of genetically engineered organisms. These publications are considered worldwide as sustainable references for use in biosafety evaluation.

This document deals with the environmental risk/safety assessment of transgenic plants at a broad level. Its purpose is to describe an approach and provide illustrative examples for planning and structuring risk/safety assessments for the release of transgenic plants into the environment. It provides general information on key concepts and important points that risk/safety assessors should focus on when planning such assessments. These key features include the comparative approach, the familiarity with the biology of the unmodified plant species, the general protection goals, the assessment endpoints, the potential adverse effects associated with the environmental release, the pathways to harm and corresponding risk hypotheses, relevant information elements, and the use of environmental considerations in planning such assessment.

Annexes A to G of the document describe seven examples of environmental considerations routinely examined by assessors and taken from actual experience gained during risk/safety assessment of transgenic plants intended for environmental release. These environmental considerations are: Invasiveness and weediness; Vertical gene flow; Organisms (animals); Soil functions; Plant health; Crop management practices; and Biodiversity (protected species and habitats/ecosystems).

The set of science-based information and data contained in this volume, previously agreed by consensus and published by the OECD, constitutes a solid reference and a practical tool for use during the biosafety assessment planning process. This publication should be of interest to regulators and assessors from national authorities in charge of evaluating the risk/safety of transgenic plants prior to environmental release, as well as to plant breeders and the wider scientific community.

1 Environmental considerations for risk/safety assessment for the release of transgenic plants

This document deals with the environmental risk/safety assessment (biosafety) at a broad level. It provides general information on key concepts and points that risk/safety assessors should focus on when planning risk/safety assessments for the release of transgenic plants into the environment: comparative approach, familiarity with the biology of the unmodified plant species, general protection goals, assessment endpoints, potential adverse effects associated with the environmental release, pathways to harm and corresponding risk hypotheses, information elements, and the use of environmental considerations in planning such assessment. Annexes A to G describe seven examples of environmental considerations routinely examined and taken from actual experience gained during risk/safety assessment of transgenic plants intended for environmental release.

1.1. Purpose of the document

The purpose of this document is to illustrate how a set of environmental considerations can be used to inform the planning and structure of an environmental risk/safety assessment for release of a transgenic plant.

The environmental considerations included in the annexes to this document were drawn from the collective knowledge of risk/safety assessors with experience in evaluating the environmental risk/safety of transgenic plants. This set of considerations captures many of the types of interactions that can occur between a transgenic plant and its receiving environment that are widely considered under various national and regional legal frameworks. They include invasiveness and weediness, vertical gene flow, organisms (animals), soil functions, plant health, crop management practices, and biodiversity (protected species and habitats/ecosystems). This set is not meant to be prescriptive or exhaustive and these considerations may be treated differently among jurisdictions.

For each environmental consideration listed above, Annexes A-G give examples of how the approach described in the next section, Planning an Environmental Risk/Safety Assessment, can facilitate the development of plausible pathways to harm to an environmental value to be protected, the formulation of corresponding risk hypotheses, and the identification of information relevant to evaluate those hypotheses. Conducting the subsequent environmental risk/safety assessment is not covered in this document. The paradigm of risk assessment has been elaborated in an earlier OECD document (OECD, 1993).

Key concepts and terms are described, including for each environmental consideration, but in some cases, they may be defined slightly differently dependent on the context, including authorship, scientific field, or jurisdiction. Even within the environmental risk/safety assessment literature (e.g. OECD, 2003; IPCS, 2004; EFSA, 2012), depending on the type of assessment, legislation, jurisdictional, or institutional framework, etc., different terms may be used to describe similar concepts. However, the purpose of this document is not to elaborate on or establish new terminology but rather to describe a process and provide illustrative examples for planning an environmental risk/safety assessment.

This document builds on the work begun by the OECD that first articulated some of the key concepts that form the context and basis for conducting an environmental risk/safety assessment: *Recombinant DNA Safety Considerations* (the so-called "Blue Book"; OECD, 1986), *Safety Considerations for Biotechnology* (OECD, 1992), and *Safety Considerations for Biotechnology: Scale-up of Crop Plants* (OECD, 1993). These concepts, which have been adopted and articulated elsewhere (UNEP, 1995; SCBD, 2000; IPPC, 2019), include: the step-by-step approach to environmental release; the case-by-case and comparative nature of the assessment; the importance of familiarity; and consideration of the characteristics of the organism, the introduced trait, the receiving environment, and the interactions among them.

Companion documents prepared by the OECD Working Party on the Harmonisation of Regulatory Oversight in Biotechnology provide additional support for the environmental risk/safety assessment of transgenic plants. These include:

- Consensus Document on Molecular Characterisation of Plants Derived from Modern Biotechnology (OECD, 2010)
- Revised Points to Consider for Consensus Documents on the Biology of Cultivated Plants (OECD, 2020)
- Series of biology of plants consensus documents (OECD, 1997 to 2021+)
- Series of trait consensus documents (OECD, 1996 to 2007+)

Taken together, this comprehensive package of OECD documents is intended to inform those conducting environmental risk/safety assessments of transgenic plants, support OECD collaborations and discussions, provide key documents both for countries that have established regulatory systems and for those establishing regulatory systems and capabilities, and inform other interested stakeholders.

1.2. Planning an environmental risk/safety assessment

Key concepts used throughout the environmental risk/safety assessment of a transgenic plant include the comparative nature of the assessment, and familiarity with the characteristics of the plant species, the introduced trait, the receiving environment, and the interactions among them. Furthermore, the assessor abides by the relevant national or regional legislation.

1.2.1. Comparative approach

Environmental risk/safety assessments typically use a comparative approach. The differences between a particular transgenic plant and a comparator provide a starting point for determining if the release of the transgenic plant might result in potential adverse effects on the environment. The transgenic plant is typically compared to a non-modified plant with a genotype that is as closely related as possible to the transgenic plant. However, there is no single concept of an appropriate comparator that is agreed upon internationally. In some instances, where the regulatory framework permits, the comparator may be another transgenic plant. Furthermore, more than one comparator may be used in a risk/safety assessment (though for simplicity, in this document, the singular 'comparator' is used). The choice of comparator can depend on the scientific questions to be considered and other factors, such as the availability of appropriate comparators and specific regulatory requirements.

When a relevant difference is identified between the transgenic plant and a comparator, it is evaluated to determine if it is significant and has biological relevance related to a jurisdiction's protection goals (see below). The variation within cultivated varieties of the plant species is usually considered to put any identified differences between the transgenic plant and the comparator into context.

1.2.2. Familiarity

Familiarity arises from knowledge of and experience with the biology of the unmodified plant, the introduced trait, and the receiving environment (OECD 1993), and plays a key role in setting the context for the environmental risk/safety assessment.

Familiarity with the plant might derive from, but is not limited to, knowledge of the plant's taxonomy and genetics, morphological characteristics, and reproductive biology. For additional information, see Revised Points to Consider for Consensus Documents on the Biology of Cultivated Plants (OECD, 2020).

Familiarity with the introduced trait might derive from, but is not limited to, knowledge of the function of the DNA sequence in its source organism, the function of the DNA sequence in the transgenic plant, and the resulting phenotype of the transgenic plant.

Familiarity with the receiving environment might derive from, but is not limited to, knowledge of the habitats available to the transgenic plant, presence and habitats of sexually-compatible species including wild relatives, centre(s) of origin and distribution, presence of species of conservation concern, provision of ecological functions, climate, growing season, presence of abiotic and biotic stressors, and types of crop management practices used, among others. The receiving environment can differ between and within regions. Therefore, consideration is given to the region where the transgenic plant will be cultivated or could reasonably be expected to grow.

The receiving environment to be considered could include both managed and unmanaged ecosystems, depending on a jurisdiction's legislative framework. There are no internationally-agreed definitions for managed and unmanaged ecosystems so for the purpose of this document, managed ecosystems are considered to include production areas for agriculture (including field margins), horticulture, and forestry, and intensive land use areas such as roadsides and urban areas. Unmanaged ecosystems include natural areas, protected reserves and parks, and other areas with minimal human intervention.

During the initial steps of the environmental risk/safety assessment of a transgenic plant intended for release, the assessor plans how to proceed with the assessment. The initial steps of the assessment can build from what is often referred to as problem formulation in the environmental risk assessment literature (Suter, 2007; Wolt et al., 2010) and include the following steps:

- Identifying general (and, when needed, operational) protection goals.
- Determining assessment endpoints.
- Identifying potential adverse effects on the assessment endpoints associated with the release of the transgenic plant.
- Identifying plausible pathways to harm to the assessment endpoints and formulating corresponding risk hypotheses for each step of the pathway.
- Determining information elements relevant to evaluating the risk hypotheses.

Subsequent steps in the risk/safety assessment, for example collecting appropriate information and data to establish the validity of the risk hypotheses identified in the planning stage, risk characterisation and decision making, fall outside the scope of this document.

1.2.3. General protection goals

General protection goals establish the context for the environmental risk/safety assessment. They describe components of the environment (e.g. species, habitats, services, etc.) that are generally identified in the relevant existing laws or policies of a jurisdiction as valued and/or protected. Specific components of the environment may be valued for their aesthetic, cultural or intrinsic value, or because they are explicitly protected by law (e.g. organisms classified as threatened or endangered). General protection goals cover broad concepts and are similar between regulatory authorities, although they may be described using different terminology. An example of a general protection goal relevant to the environmental risk/safety assessment of a transgenic plant could be 'sustainability of ecosystem services'.

1.2.4. Assessment endpoints

Assessment endpoints are derived from general protection goals and are explicit expressions of the environmental value to be protected. Assessment endpoints can be further defined as a valued ecological entity and an attribute that can be estimated by measurement or modelling. When general protection goals are too broad to translate directly into assessment endpoints, operational protection goals derived from the general protection goals may be used as an intermediate step to facilitate the determination of assessment endpoints (Garcia-Alonso and Raybould, 2014; Devos et al., 2015). For example, the general protection goal 'sustainability of ecosystem services' could be refined into several operational protection goals, notably 'maintaining pollination services', then further into assessment endpoints, each consisting of an entity and an attribute. The entity for 'maintaining pollination services' could be pollinators/honeybees, and its attributes could be at organism-level (e.g. pollinator/honeybee survival) and/or at population-level (e.g. pollinator/honeybee abundance) (U.S. EPA, 2007). It should be noted that in an environmental risk/safety assessment it is often necessary to consider a number of assessment endpoints to address each general or operational protection goal.

1.2.5. Potential adverse effects associated with the environmental release of a transgenic plant

If a transgenic plant is released, its interactions with the receiving environment may or may not adversely affect an assessment endpoint. The identification of potential adverse effects may be informed by characteristics of the plant, trait and receiving environment. For example, if pollinator/honeybee survival is selected as an assessment endpoint, a potential adverse effect that could be postulated for assessment is that the trait might affect pollinator/honeybee survival leading to reduced abundance and a reduction in pollination services.

1.2.6. Pathways to harm and corresponding risk hypotheses

Pathways to harm (causal or conditional chains of events) describe the scientifically plausible and necessary steps that would need to occur for release of the particular transgenic plant to result in an adverse effect on the assessment endpoint (Nickson, 2008). When planning the environmental risk/safety assessment, one or more pathways leading to harm may be postulated by the assessor for each potential adverse effect identified for an assessment endpoint.

The simple linear examples of pathways provided in the annexes of this document are for illustrative purposes. In reality, the process is often more complex. For example, there may be more than one plausible pathway to consider when determining whether an assessment endpoint may be adversely affected by the interaction of a transgenic plant with its receiving environment. In addition, multiple, plausible pathways may share some of the same steps.

For each step of a postulated pathway to harm, a corresponding risk hypothesis is formulated that will enable the risk assessor to determine whether the pathway is likely to occur. A risk hypothesis can be evaluated in a number of ways that include but are not limited to using experimental data or information available from the scientific literature, or other relevant information as deemed appropriate by the risk assessor (e.g. climate or herbarium studies). If in the actual environmental risk/safety assessment, the evaluation of a risk hypothesis concludes that a step in a pathway is unlikely to occur, then the likelihood of the harm occurring through that particular pathway most likely is negligible. In practice, some hypotheses may be difficult to evaluate or the evaluation using available information may not produce definitive conclusions regarding the likelihood of a particular step in a pathway. This uncertainty may be addressed through a tier-based testing approach (U.S. EPA, 2007), by consideration of multiple sources of information and lines of evidence (i.e. a weight of evidence approach), or by new studies being undertaken (Devos et al., 2019). Nevertheless, in some cases uncertainties may remain that must be addressed by decision makers and risk managers.

1.2.7. Information elements

Information elements that provide the evidence to evaluate the validity of each risk hypothesis are identified. Such evidence can be obtained from a variety of sources as indicated in the previous paragraph. Information elements may relate to characteristics of the plant, the trait, or the receiving environment, and may be quantitative or qualitative. A single information element may be relevant for the evaluation of multiple risk hypotheses. Information elements are only relevant to the assessment when they address a particular risk hypothesis (Devos et al., 2019).

1.3. Use of environmental considerations in planning an environmental risk/safety assessment

Annexes A-G describe a set of environmental considerations routinely examined by assessors when carrying out risk/safety assessments of transgenic plants intended for environmental release. This set includes:

- **Annex A. Invasiveness and Weediness**: This annex provides illustrative examples when considering whether a transgenic plant has the potential to have adverse effects on the environment due to increased weediness or invasiveness, relative to the comparator.
- **Annex B. Vertical Gene Flow**: This annex provides illustrative examples when considering whether gene flow from a transgenic plant to sexually-compatible plants (weedy relatives, and valued relatives and landraces) might represent an additional, indirect pathway of exposure of the environment to the transgenic plant. Gene flow is not an adverse effect per se, but its consequences may lead to adverse environmental effects, relative to the comparator.
- **Annex C. Organisms (Animals)**: This annex provides illustrative examples when considering whether a transgenic plant has the potential to i) have adverse effects on organisms in the environment and their role in ecological functions including food webs, relative to the comparator, and ii) have adverse effects on human/animal health due to non-dietary exposure, relative to the comparator.
- **Annex D. Soil Functions**: This annex provides illustrative examples when considering whether a transgenic plant has the potential to have adverse effects on soil microbial communities responsible for soil processes and soil functions, relative to the comparator.
- **Annex E. Plant Health**: This annex provides illustrative examples when considering whether a transgenic plant has the potential to have adverse effects on its health and the health of surrounding plants in the environment by having an enhanced ability to act as a host for pests, relative to the comparator.
- **Annex F. Crop Management Practices**: This annex provides illustrative examples when considering whether use of a transgenic plant has the potential to drive changes in crop management practices associated with its cultivation relative to those associated with the cultivation of the comparator, and whether such changes could have adverse effects on the environment.
- **Annex G. Biodiversity (Protected Species and Habitats/Ecosystems)**: This annex provides illustrative examples when considering whether a transgenic plant has the potential to have adverse effects on species and habitats explicitly protected by legislation of a country or a region, relative to the comparator, while the six previously- mentioned annexes provide examples when considering a selection of ways in which a transgenic plant has the potential to have adverse effects on species or ecosystems that may have a role in ecological functions and services. Broader aspects of biodiversity may be addressed in an environmental risk assessment for the release of transgenic plant under some jurisdictions.

Each annex is organised in the following manner:

- An introduction describing the environmental consideration.
- Key concepts and terms relevant to the environmental consideration.
- Determination of assessment endpoints.
- Identification of potential adverse effects on the assessment endpoints.
- Identification of plausible pathways to harm, formulation of risk hypotheses, and examples of information elements relevant to the risk hypotheses.

The set of environmental considerations included in the annexes captures many of the types of interactions that can occur between plants (including transgenic plants) and their receiving environments. They are meant to provide a convenient way of planning an environmental risk/safety assessment of a transgenic plant based on the nature of potential biological interactions of a plant with its environment. This set is not taken from any single country's considerations or terminology but is reflective of the aspects widely considered by various countries or regions under their legislative frameworks.

Not all environmental considerations may apply in each risk/safety assessment and those that do apply in a particular case will depend on the characteristics of the plant, trait, and receiving environment, and the interactions amongst them. The relevance of each consideration may also vary based on jurisdictional regulatory schemes and general protection goals. Relevant environmental considerations may be addressed in any order that is appropriate to the environmental risk/safety assessment being planned.

As the document is intended to be illustrative rather than comprehensive, only one or two examples are given for each environmental consideration on how the described approach can be used to plan the environmental risk/safety assessment of a particular transgenic plant. The examples are taken from actual experience gained during risk/safety assessment of transgenic plants that have already been evaluated somewhere in the world including, in particular, herbicide-tolerant and/or insect-resistant maize, cotton, low-erucic acid rapeseed (canola), and soybean. The approach described in this document may be considered for different parental plant types, for traits with less familiarity, or in other situations where a plant may be subject to an environmental risk/safety assessment.

References

Devos, Y. et al. (2019), "Using problem formulation for fit-for-purpose pre-market environmental risk assessments of regulated stressors", *EFSA Journal*, Vol. 17 (51), pp. 31, https://doi.org/10.2903/j.efsa.2019.e170708.

Devos, Y. et al. (2015), "Optimising environmental risk assessments – Accounting for biodiversity and ecosystem services helps to translate broad policy protection goals into specific operational ones for environmental risk assessments", *EMBO reports*, Vol. 16 (9), pp. 1060-1063, https://doi.org/10.15252/embr.201540874.

EFSA Scientific Committee (2012), "Scientific Opinion on Risk Assessment Terminology", *EFSA Journal*, Vol. 10 (5): 2664, pp.43, https://doi.org/10.2903/j.efsa.2012.2664.

Garcia-Alonso, M. and A. Raybould (2014), "Protection goals in environmental risk assessment: a practical approach", *Transgenic Research*, Vol. 23 (6), pp. 945-956, https://doi.org/10.1007/s11248-013-9760-1.

IPCS (2004), *IPCS Risk Assessment Terminology*, International Programme on Chemical Safety, Inter-organisation Programme for the Sound Management of Chemicals, World Health Organization, http://www.inchem.org/documents/harmproj/harmproj/harmproj1.pdf (accessed 27 April 2023).

IPPC (2019), *Pest risk analysis for quarantine pests - ISPM 11* [adopted 2013; published 2019], International Standards for Phytosanitary Measures, International Plant Protection Convention, Food and Agriculture Organization of the United Nations, ISPM 11. Pest risk analysis for quarantine pests (fao.org).

Nickson T. E. (2008), "Planning environmental risk assessment for genetically modified crops: problem formulation for stress-tolerant crops", *Plant Physiology*, Vol. 147 (2), pp. 494–502. https://doi.org/10.1104/pp.108.118422.

OECD (2020), “Revised Points to Consider for Consensus Documents on the Biology of Cultivated Plants”, *Series on Harmonisation of Regulatory Oversight in Biotechnology, No.67,* https://one.oecd.org/document/env/jm/mono(2020)2/en/pdf.

OECD (1997 to 2021+), “Consensus Documents for the Work on Harmonisation of Regulatory Oversight in Biotechnology: Biology of Plants”, *Series on Harmonisation of Regulatory Oversight in Biotechnology,* http://www.oecd.org/env/ehs/biotrack/consensusdocumentsfortheworkonharmonisationofregulatoryoversightinbiotechnologybiologyofcrops.htm (accessed 27 April 2023).

OECD (2010), “Consensus Document on Molecular Characterisation of Plants Derived from Modern Biotechnology”, *Series on Harmonisation of Regulatory Oversight in Biotechnology, No.51,* https://www.oecd.org/science/biotrack/46815346.pdf (accessed 27 April 2023).

OECD (1996 to 2007+), “Consensus Documents for the Work on Harmonisation of Regulatory Oversight in Biotechnology: Traits”, *Series on Harmonisation of Regulatory Oversight in Biotechnology,* http://www.oecd.org/env/ehs/biotrack/consensusdocumentsfortheworkonharmonisationofregulatoryoversightinbiotechnologytraits.htm (accessed 27 April 2023).

OECD (2003), “Descriptions of selected key generic terms used in chemical hazard/risk assessment”, *Series on Testing and Assessment, No. 44,* http://www.oecd.org/officialdocuments/publicdisplaydocumentpdf/?cote=ENV/JM/MONO(2003)15&docLanguage=En (accessed 27 April 2023).

OECD (1993), *Safety Considerations for Biotechnology: Scale-up of Crop Plants*, Organisation for Economic Cooperation and Development, http://www.oecd.org/env/ehs/biotrack/1958527.pdf (accessed 27 April 2023).

OECD (1992), *Safety Considerations for Biotechnology*, Organisation for Economic Cooperation and Development, http://www.oecd.org/sti/emerging-tech/2375496.pdf (accessed 27 April 2023).

OECD (1986), Recombinant DNA Safety Considerations – Safety Considerations of industrial, agricultural and environmental applications of organisms derived by recombinant DNA techniques, Organisation for Economic Cooperation and Development, http://www.oecd.org/env/ehs/biotrack/Recombinant-DNA-Safety-Considerations.pdf (accessed 27 April 2023).

SCBD (2000), Cartagena Protocol on Biosafety to the convention on biological diversity: text and annexes, Secretariat of the Convention on Biological Diversity, https://www.cbd.int/doc/legal/cartagena-protocol-en.pdf (accessed 27 April 2023).

Suter II, G.W. (2006), Ecological Risk Assessment, Second Edition, CRC Press, Taylor and Francis Group, https://doi.org/10.1201/9781420012569.

UNEP (1995), UNEP International technical guidelines for safety in biotechnology, United Nations Environment Programme. https://digitallibrary.un.org/record/222403 (accessed 27 April 2023).

U.S. EPA and USDA (2007), White Paper on Tier-Based Testing for the Effects of Proteinaceous Insecticidal Plant-Incorporated Protectants on Non-Target Arthropods for Regulatory Risk Assessments, United States Department of Agriculture and United States Environmental Protection Agency, https://www.epa.gov/sites/production/files/2015-09/documents/tier-based-testing.pdf (accessed 27 April 2023).

Wolt, J.D. et al. (2010), “Problem formulation in the environmental risk assessment for genetically modified plants”, Transgenic Research, Vol. 19, pp. 425-436, https://doi.org/10.1007/s11248-009-9321-9.

Annex A. Invasiveness and Weediness

A consideration for the environmental risk/safety assessment of a transgenic plant is the evaluation of its potential to have adverse effects on the environment due to increased invasiveness or weediness, relative to the comparator.

Concepts and terms

Whether a particular plant is considered an invasive plant or weed can be context dependent in terms of where and when it emerges and what impact it has on the environment. A cultivated plant may be valued in an urban, agricultural or pastoral setting but may be invasive of natural environments (e.g. olive trees in Australia). A naturally occurring plant may be valued in unmanaged ecosystems but be considered a weed if it emerges in crops and competes for resources[1] (e.g. in North America, milkweed is essential to the Monarch butterfly in unmanaged ecosystems but considered a weed in maize crops). Many documented invasive plants are exotic species that have been introduced to managed and unmanaged ecosystems, accidentally or deliberately, as a result of human cultivation (including ornamentals, pasture species, trees, crops) or human mediated transport (see Pysek et al., 2017, Randall et al., 2017).

It should be noted that there are varied definitions and usages of 'invasive plant', 'invasiveness' or 'weed', 'weedy', 'weediness' (e.g. see Richardson et al., 2011; Lockwood et al., 2013). While various terminologies may make distinctions between weeds and invasive plants, these terms may be considered broadly synonymous and imply the potential for adverse effects on the environment.

Which terms are used depends on context, including in different jurisdictions and legal frameworks, international organisations (e.g. see IPPC, 2019; IPPC, 2007), scientific disciplines (e.g. weed science vs. invasion biology) and fields of human activity (e.g. agriculture vs environmental protection vs. urban management). The term weed (and weed control) is often used in the context of agriculture or other managed ecosystems but it is also used as a generic descriptor for any plant considered to be a problem, a pest plant, or occurring where it is not wanted. The term invasive plant (or invasive species) is often used in the context of unmanaged ecosystems, though the term weed is also used in this environmental context in some disciplines and jurisdictions. While not all weeds are invasive, this environmental consideration focuses on the impact of the invasiveness of weeds in the environment.

For simplicity, this document uses the terms invasiveness and invasive potential rather than making distinctions between weediness and invasiveness.

Invasion of an ecosystem, whether by plants or other organisms, is considered to be a staged process with a series of conceptual steps (e.g. Blackburn et al., 2011): introduction or entry to the ecosystem; establishment; persistence and survival to maturity; reproduction; dispersal of propagules; increased abundance and geographic spread leading to adverse effects on the environment (e.g. by competition with valued species).

The invasive potential of a plant will depend on the characteristics of the plant and of the receiving environment. A range of characteristics related to invasiveness (derived from known invasive plants) have been proposed (e.g. *Baker's List*, Baker, 1965) and elaborated (e.g. Richardson et al., 2000), including in weed risk assessment protocols (Pheloung et al., 1999; Downey et al., 2010; FAO, 2011).

Such protocols have also been proposed for assessment of the invasive potential of transgenic plants (Kos et al., 2011; Keese et al., 2014). Whether a given plant will be invasive is dependent on multiple factors related to the characteristics of the plant and the receiving environment. The presence of one or more 'invasive characteristics' is not determinative.

Key plant characteristics contributing to invasive potential include those that confer the ability to:

- Establish (e.g. be introduced, germinate, grow) and persist (e.g. survive over time) in a receiving environment;
- Reproduce, generally quickly with large numbers of progeny (e.g. short time to sexual maturity and seed set, large numbers of propagules, including asexual propagules);
- Disperse, generally in large numbers (e.g. small seed size, pod shattering, vegetative propagules or spread over long distances);
- Compete well in the environment (e.g. rapid growth, suppressive growth habit); and
- Overcome abiotic (e.g. climatic conditions), biotic (e.g. competitors, herbivores, diseases) or human (e.g. weed control) constraints.

These characteristics are incorporated in OECD biology documents as described in the *Revised Points to Consider for Consensus Documents on the Biology of Cultivated Plants* (OECD, 2020) and pest risk analysis guidance (e.g. IPPC, 2019). Many domesticated plants have lost invasiveness traits through breeding (Kos et al., 2011). Crop or other plants described as weedy possess or display some traits associated with invasiveness but may not be considered invasive.

Persistence is the ability of plants or their progeny, plant reproductive propagules (seed or vegetative propagules such as rhizomes), to survive (i.e. remain viable), and/or reach maturity and reproduce, and/or remain dormant across growing seasons (e.g. ability of seeds, propagules to enter dormancy and survive in soil over time).

Spread/dispersal is the ability of plants or plant reproductive or vegetative propagules (e.g. ivy stems can establish new plants) to move in the environment by natural (e.g. creeping growth habit or subterranean runners) or human-assisted means (e.g. transport of seed, vegetative propagules after harvest). Many plants and plant propagules have structures or characteristics that facilitate spread directly (e.g. pod shattering), by wind (e.g. seed with fluffy outgrowths, tumbleweeds), by water (e.g. buoyant, water resistant, seeds with shells, air pockets or larger surface area), or by animals or human activity (e.g. hooks or spines that attach seeds to animals or objects (such as machinery), or fleshy fruit attractive to animals such as birds).

Problem formulation

For this consideration, below is a simple example that illustrates the approach for planning an environmental risk/safety assessment. It includes a discussion of assessment endpoints, potential adverse effects, and a linear pathway to harm with corresponding risk hypotheses and information elements to illustrate the approach. As previously indicated in the document (section 1.2.6), the process is often more complex.

(a) Determination of assessment endpoints

As with any cultivated plant, a transgenic plant might have impacts on the environment if it is invasive. An example of an assessment endpoint relevant to evaluating whether the transgenic plant has increased invasive potential relative to the comparator is the abundance of a valued plant species in unmanaged ecosystems.

(b) Identification of potential adverse effects on the assessment endpoints

The identification of potential adverse effects from invasiveness of a transgenic plant should be informed by the characteristics of the comparator (e.g. the unmodified plant), of the transgenic plant (trait, phenotype), and of the potential receiving environment(s). Adverse effects from (non-transgenic) invasive plants often result from competition with and displacement of other plants in managed and/or unmanaged ecosystems.

Knowledge of the invasive potential of the comparator provides important contextual baseline information, for example, whether the plant has any invasive characteristics, whether it survives outside of human cultivation, whether it is considered invasive and where it invades (e.g. in managed or unmanaged ecosystems), and what adverse effects it causes.

Knowledge of the potential receiving environment(s) provides information on where and what adverse effects might occur. For example, if the environment is suitable to support establishment, persistence and growth of the comparator or transgenic plant, whether and how those ecosystems are susceptible or resistant to invasion, whether human management activities are controlling or mitigating invasive impacts, or whether the receiving environment has conditions that might be more conducive to the survival or spread of the transgenic plant than the comparator.

The nature of the trait and phenotype of the transgenic plant informs identification of potential adverse effects, especially whether and how they could increase the invasive potential of the transgenic plant relative to the comparator. Consideration of the transgenic trait will indicate if it may confer or enhance invasiveness traits, or enable the transgenic plant to overcome natural or human constraints that limit the comparator (i.e. confer a competitive fitness advantage). If the trait confers tolerance to water stress, it might enable the transgenic plant to better establish and persist outside of crop fields (i.e. without human irrigation). If the trait confers tolerance to an herbicide, the transgenic plant might escape current controls with that herbicide.

If the transgenic plant has increased invasive potential relative to the comparator, it may establish in ecosystems outside the fields in which it is cultivated. An example of a potential adverse effect on the environment from increased invasive potential according to the assessment endpoints identified above may include reduced abundance of a valued plant species in unmanaged ecosystems.

(c) Identification of plausible pathways to harm, formulation of risk hypotheses, and identification of information elements relevant to evaluating the risk hypotheses

In this section, a plausible pathway to harm is postulated. For each step of the postulated pathway to harm, a corresponding risk hypothesis is formulated that will enable the risk assessor to determine whether the pathway is likely to occur. Once it is shown that any part of the pathway is highly unlikely, one does not need to continue evaluating the subsequent steps in the pathway and can conclude that the specific pathway to harm is unlikely to occur. In addition, examples of information elements that can be used to evaluate the risk hypotheses are given along with their rationales.

Postulated pathway leading to reduced abundance of a valued plant species in unmanaged ecosystems

Propagules (e.g. seeds) from crops can spread from cultivated fields by a variety of means and sometimes establish and persist in unmanaged ecosystems. If a transgenic plant has a relevant changed phenotype (e.g. increased tolerance to an abiotic stressor such as drought), it might have an increased ability, relative to the comparator, to establish, persist, reproduce and spread in unmanaged ecosystems, to compete with other plant species, and thus lead to a reduction in the abundance of a valued plant species in unmanaged ecosystems.

One example of a postulated pathway to harm for this adverse effect is shown in the first column of Table A A.1. Risk hypotheses for each step of the pathway are formulated in the second column and the third column provides examples of information elements for evaluating the hypotheses.

Table A A.1. Postulated pathway leading to reduced abundance of valued plant species in unmanaged ecosystems due to increased invasive potential of the transgenic plant, corresponding risk hypotheses, and relevant information elements

Pathway steps	Risk hypotheses	Examples of information elements
The introduced trait in the transgenic plant confers increased drought-tolerance relative to the comparator	The introduced trait in the transgenic plant does not confer increased drought-tolerance relative to the comparator	Knowledge about the comparator and the nature of the trait and phenotype of the transgenic plant, including the level of drought-tolerance under water stress conditions
Propagules of the transgenic plant are introduced into unmanaged ecosystems	Propagules of the transgenic plant are not introduced into unmanaged ecosystems	Cultivation, harvest and transport practices for the crop, proximity of cropping areas to the unmanaged ecosystem; Biology of the comparator, including propagule dispersal characteristics
Propagules of the transgenic plant germinate, establish and persist in unmanaged ecosystems	Propagules of the transgenic plant do not germinate, establish and persist in unmanaged ecosystems	Reproductive biology of the comparator and transgenic plant, including propagule dormancy; Characteristics and climate of the unmanaged ecosystem e.g. occurrence of drought/rainfall
The transgenic plant reproduces and spreads, resulting in increased abundance in unmanaged ecosystems, relative to the comparator (i.e. the transgenic plant has a fitness advantage and increased invasive potential)	The transgenic plant does not reproduce, spread or increase in abundance in unmanaged ecosystems, relative to the comparator (i.e. the transgenic plant does not have a fitness advantage or increased invasive potential)	Biology of the comparator including any previous history of invasiveness; Data collected on the transgenic plant relative to the comparator (e.g. propagule numbers, numbers or biomass of plants establishing and reproducing under water stress)
The abundance of a valued plant species in unmanaged ecosystems is reduced due to competition from the transgenic plant	The abundance of a valued plant species in unmanaged ecosystems is not reduced due to competition from the transgenic plant	Biology of the valued plant species and ecology of the unmanaged ecosystem; History of invasion of the unmanaged ecosystem
The abundance of valued plant species in unmanaged ecosystems is reduced		

It is important to note that examples of information elements in this table are intended to illustrate the types of information that can be used in evaluating a risk hypothesis, i.e. to determine whether particular pathway steps are likely to occur. However, for any step there might be other information that would be relevant. Rationales for how such information elements may be used to evaluate the risk hypotheses include:

- The level of drought-tolerance of the comparator and the transgenic plant provides information on the potential for increased survival of the transgenic plant in drought conditions, including drought conditions that occur in the unmanaged ecosystem. The level of drought-tolerance is relevant to multiple steps in the pathway;
- Knowledge of the cultivation, harvest and transport practices for the crop, including the proximity of the cropping area to the unmanaged ecosystem, provides information about potential routes for introduction of the transgenic plant to the unmanaged ecosystem;

- The propagule characteristics of the comparator (e.g. structures that facilitate dispersal by wind, water, animals, machinery) provide information on how propagules of the transgenic plant might be introduced to the unmanaged ecosystem;
- Propagule dormancy characteristics provide information on the ability of the comparator or transgenic plant to persist over multiple seasons (e.g. from a seed bank);
- Knowledge of the ecology of the unmanaged ecosystem, including the occurrence of drought (e.g. from rainfall records), provides information on whether the drought-tolerant trait may increase the survival of the transgenic plant in that ecosystem;
- The invasive history of the comparator, either locally or elsewhere in the world, provides information on the invasive potential of the species, including whether drought is a limiting factor. This type of information might include whether the comparator is already present in the unmanaged ecosystem, or whether the species has been identified by the relevant jurisdiction as undesirable (e.g. a 'noxious weed' or a, 'pest'[2]);
- The reproductive and dispersal biology of the comparator (e.g. pollination requirements, numbers of propagules, and dispersal mechanisms, such as wind or pod shattering) provides information on how the transgenic plant may reproduce, spread and increase in abundance in the unmanaged ecosystem;
- The biology and ecology of the valued plant species, e.g. its abundance and distribution in the unmanaged ecosystem, provide information on whether and how it might be affected by the abundance of the transgenic plant (e.g. do they share an ecological niche?). The level of drought-tolerance of the valued plant species provides information on whether the transgenic plant would have a competitive advantage in drought conditions;
- The history of invasion of the unmanaged ecosystem may provide information on whether that ecosystem is susceptible or resistant to plant invasions, and what factors (e.g. drought-tolerance) were important in any previous invasions.

References

Baker, H.G. (1965), "Characteristics and modes of origin of weeds", in Baker, H.G. and G.L. Stebbins (eds.), *The genetics of colonizing species*, pp. 147-172, Academic Press, NY.

Blackburn, T.M. et al. (2011), "A proposed unified framework for biological invasions", *Trends in Ecology and Evolution*, Vol. 26, pp. 332-339, https://doi.org/10.1016/j.tree.2011.03.023.

Downey, P.O. et al. (2010), "Assessing risk across the spectrum of weed management", *CAB Reviews: Perspectives in Agriculture, Veterinary Science, Nutrition and Natural Resources*, Vol. 5, (38), pp. 1-15, https://doi.org/10.1079/PAVSNNR20105038.

FAO (2011), *Procedures for Post-border Weed Risk Management, 2nd Edition*, Plant Production and Protection Division, Food and Agriculture Organization of the United Nations, https://www.fao.org/fileadmin/templates/agphome/documents/Biodiversity-pollination/Weeds/Docs/ProceduresPostBorderWeedRiskManagement.pdf.

IPPC (2019), *Pest risk analysis for quarantine pests - ISPM 11* [adopted 2013; published 2019], International Standards for Phytosanitary Measures, International Plant Protection Convention, Food and Agriculture Organization of the United Nations, ISPM 11. Pest risk analysis for quarantine pests (fao.org).

IPPC (2007), *Glossary of Phytosanitary Terms - ISPM No. 5,* International Standards for Phytosanitary Measures, International Plant Protection Convention, Food and Agriculture Organization of the United Nations, https://www.ippc.int/largefiles/adopted_ISPMs_previousversions/en/ISPM_05_2007_En_2007-07-26.pdf.

Keese, P.K. et al. (2014), "Applying a weed risk assessment approach to GM crops", *Transgenic Research*, Vol. 23, pp. 957–969, https://doi.org/10.1007/s11248-013-9745-0.

Kos, S.P., T.J. De Jong and W.L.M. Tamis (2011), "Can transgenic crops go wild? A literature study on using plant traits for weediness pre-screening", *COGEM report*, CGM 2012-01, https://cogem.net/app/uploads/2019/07/CGM-2012-01-Weediness-pre-screening1.pdf.

Lockwood, J.L., M.F. Hoopes and M.P. Marchetti (2013), *Invasion Ecology, 2nd Edition*, Wiley-Blackwell, Chichester, West Sussex, UK.

OECD (2020), "Revised Points to Consider for Consensus Documents on the Biology of Cultivated Plants", *Series on Harmonisation of Regulatory Oversight in Biotechnology, No.67,* https://one.oecd.org/document/env/jm/mono(2020)2/en/pdf.

Pheloung, P.C., P.A. Williams and S.R. Halloy (1999), "A weed risk assessment model for use as a biosecurity tool evaluating plant introductions", *Journal of Environmental Management*, Vol. 57, pp. 239–251, https://doi.org/10.1006/jema.1999.0297.

Pysek, P. et al. (2017), "Naturalized alien flora of the world: species diversity, taxonomic and phylogenetic patterns, geographic distribution and global hotspots of plant invasion", *Preslia*, Vol. 89, pp. 203–274, https://www.preslia.cz/article/view?id=60.

Randall, R.P. (2017), *A Global Compendium of Weeds. 3rd Edition*, Perth, Western Australia, R.P. Randall, https://www.researchgate.net/publication/313645439_A_Global_Compendium_of_Weeds_Third_Edition.

Richardson, D., P. Pyšek and J. Carlton (2011), "A Compendium of Essential Concepts and Terminology in Invasion Ecology", in *Fifty Years of Invasion Ecology: The Legacy of Charles Elton*, pp. 409-420, Blackwell Publishing Ltd, https://doi.org/10.1002/9781444329988.ch30.

Richardson, D.M. et al. (2000), "Naturalization and invasion of alien plants: concepts and definitions", *Diversity and Distributions*, Vol. 6, pp. 93-107, https://doi.org/10.1046/j.1472-4642.2000.00083.x.

Notes

[1] The impact of weeds on crop production is generally considered to be economic rather than environmental.

[2] Note that the terms used for such declarations will vary between jurisdictions and organisations.

Annex B. Vertical Gene Flow

A consideration for the environmental risk/safety assessment of a transgenic plant is the evaluation of the potential for transfer of transgenes via vertical gene flow to sexually-compatible plants to result in adverse effects on the environment, relative to the comparator. Vertical gene flow may be considered an "exposure pathway" and as such this annex differs from the other annexes in that vertical gene flow in and of itself is not an adverse effect.

This annex provides a two-step process for evaluating gene flow and its potential consequences. First, the annex provides an illustrative example to assist the assessor when considering whether transgene introgression is plausible. Second, it includes an example to assist the assessor when considering whether gene flow of a transgene, if it occurs, could have the potential to adversely affect the environment due to a change in the viability of populations of a valued species.

Concepts and terms

Vertical gene flow refers to the sexual transfer of genetic material between genetically distinct populations including the movement of genes from one population into other populations of the same species (intraspecific gene flow) or other sexually-compatible species (interspecific gene flow). Vertical gene flow is a natural process mediated by plant sexual reproduction and thus gene flow is not an adverse effect *per se*. Cultivated plant species are known to transfer genes to sexually-compatible wild relatives (Ellstrand et al., 2013).

Important steps in vertical gene flow are the spread of genetic material between donor and recipient plants, the formation of hybrids, and the stable establishment of the genetic material from the donor in the recipient population via introgression. In flowering plants, vertical gene flow is mediated by pollen, which can be dispersed by pollinators, wind, and very occasionally by water.

Introgression is the stable incorporation of genetic material (genes, alleles) in a population, generally through the repeated backcrossing of an interspecific or intraspecific hybrid with one of its parent species.

Population viability is the ability of a population to survive and persist in the environment.

Natural hybridisation involves successful mating between individuals of two genetically distinct populations or groups of populations (Harrison, 1990; Arnold, 1997). The rate of hybridisation varies between different cultivated plants and their relatives in frequency and magnitude, and mating can be uni- as well as bi-directional. Natural hybridisation is typically the first of many steps by which vertical gene flow occurs between populations (Ellstrand et al., 2013). Hybridisation may be intraspecific or interspecific.

Hybrid is the progeny from hybridisation between two genetically distinct plants.

Seed dispersal and vegetative propagation are mechanisms that plants use to spread and persist. Dispersed seed may include spatially dispersed seed from a given plant or seed from plants established via vegetative propagation. The potential for vertical gene flow can extend beyond the site where a plant was originally located/cultivated if its seed and/or vegetative propagules are spatially dispersed and establish successfully. The resulting plants may be in closer proximity to sexually-compatible relatives thus increasing the likelihood of cross-pollination.

Transgene, generally defined as a gene from a different species, is the introduced gene that confers/ determines the trait that modifies the phenotype of the transgenic plant.

Problem formulation

For this consideration, below are simple examples that illustrate the approach for planning an environmental risk/safety assessment. It includes a discussion of assessment endpoints, potential adverse effects, and a linear pathway to harm with corresponding risk hypotheses and information elements to illustrate the approach. As previously indicated in the document (section 1.2.6), the process is often more complex.

(a) Determination of assessment endpoints

Gene flow is a natural process that is common among sexually-compatible plants. Gene flow between cultivated plants, including a few transgenic plants and their sexually-compatible relatives, is well documented in the scientific literature (e.g. Kwit et al., 2011). Gene flow from a transgenic plant (i.e. the donor population) may result in the transfer of a transgene into the population of a sexually-compatible plant (i.e. the recipient population). The transgene may be permanently incorporated (introgressed) into the recipient population through several generations of hybridisation and backcrossing, especially if the transgene confers a fitness advantage. The occurrence of a hybrid progeny may lead to adverse environmental effects, depending on the trait (conferred by the transgene) under consideration.

An example of an assessment endpoint that could be affected by the occurrence of gene flow from a transgenic plant is population viability of a valued species.

(b) Identification of potential adverse effects on the assessment endpoints

As noted above, vertical gene flow is not an adverse effect *per se.* The identification of potential adverse effects on the environment resulting from vertical gene flow from a donor transgenic plant to a sexually-compatible recipient plant should be informed by the characteristics of the donor species and the trait and phenotype of the transgenic plant (conferred by the transgene), and of the potential receiving environment(s) including the characteristics of the recipient species.

If the transgenic plant is cultivated or dispersed near to a sexually-compatible plant (e.g. a weedy relative) population, interspecific hybridisation may occur and particularly if the transgene confers a fitness advantage it may be subsequently acquired by the recipient population through introgression. An example of a potential adverse effect on the environment to the assessment endpoint identified above is decreased population viability of a valued species because of increased competition from the hybrid progeny. It should be noted that depending on the trait, the types of potential adverse effects and pathways to harm detailed in the other environmental considerations annexes for transgenic plants might be relevant to such hybrid or introgressed progeny.

(c) Identification of plausible pathways to harm, formulation of risk hypotheses, and identification of information elements relevant to evaluating the risk hypotheses

In this section, a plausible pathway to harm is postulated. For each step of the postulated pathway to harm, a corresponding risk hypothesis is formulated that will enable the risk assessor to determine whether the pathway is likely to occur. Once it is shown that any part of the pathway is highly unlikely, one does not need to continue evaluating the subsequent steps in the pathway and can conclude that the specific pathway to harm is unlikely to occur. In addition, examples of information elements that can be used to evaluate the risk hypotheses are given along with their rationales.

Postulated pathway leading to decreased population viability of a valued species

Vertical gene flow may be considered as an 'exposure pathway' because, unlike most of the other environmental considerations, the focus is not on the cultivated transgenic plant itself but rather on whether recipient plants have the potential to have adverse effects on the environment.

The initial focus is therefore necessarily on whether successful hybridisation can occur between the transgenic plant and a sexually-compatible plant (e.g. weedy relative). The occurrence of gene flow is dependent on many factors, including the mating system, the degree of sexual compatibility, the life history and pollinators. The transgenic (donor) and sexually-compatible weedy relative (recipient) plants must have overlapping flowering phenology, be sufficiently close for pollination to occur, and the cross must result in viable and fertile interspecific hybrid progeny. The occurrence of introgression requires several generations of interspecific hybrids backcrossing with the recipient population.

If the transgene provides a fitness advantage in the hybrid-derived weedy population, this may increase the likelihood of introgression of the transgene into the weedy relative population. Increased competition from interspecific hybrids or introgressed plants (with a fitness advantage due to the transgene) could lead to decreased population viability of a valued species.

One example of a postulated pathway to harm for this adverse effect is shown in the first column of Table A B.1. Risk hypotheses for each step of the pathway are formulated in the second column and the third column provides examples of information elements for evaluating the hypotheses.

Table A B.1. Postulated pathway leading to gene flow occurring and decreased population viability of a valued species, corresponding risk hypotheses, and relevant information elements

Pathway steps	Risk hypotheses	Examples of information elements
The transgenic plant is cultivated within the geographic distribution range of a sexually-compatible weedy relative	The transgenic plant is not cultivated within the geographic distribution range of a sexually-compatible weedy relative	The presence of a sexually-compatible weedy relative in the receiving environment
The transgenic plant and sexually-compatible weedy relative have overlapping phenology	The transgenic plant and sexually-compatible weedy relative do not have overlapping phenology	Flowering time of the transgenic plant and sexually- compatible weedy relative within the receiving environment
The transgenic plant and sexually-compatible weedy relative hybridise in the receiving environment, producing viable and fertile transgenic progeny	The transgenic plant and sexually-compatible weedy relative do not hybridise in the receiving environment, or they do not produce viable and fertile progeny	Known hybridisation between the comparator and the sexually-compatible weedy relative and occurrence of natural hybridisation between the transgenic plant and weedy relative (e.g. indicated by a phenotypic or genotypic marker)
The vertical gene flow pathway would end here. The additional steps illustrate how transfer of the transgenic trait to a sexually-compatible weedy relative may lead to an adverse effect on the environment[1]		
The transgene has the potential to result in a fitness-advantage in the hybrid-derived weedy population	The transgene has no potential to result in a fitness-advantage in the hybrid-derived weedy population	The nature of the trait and phenotype of the transgenic plant informs identification of potential adverse effects
Presence of the transgene results in a change in fitness-associated trait(s) in the hybrid-derived weedy population	The transgene and fitness-associated trait(s) are not found in the hybrid-derived weedy population	Presence of the transgene and fitness-associated trait(s) in the hybrid-derived weedy population
The introgressed trait increases the reproductive potential of the hybrid-derived weedy relative, conferring a fitness advantage compared to the non-transgenic hybrid-derived population	The introgressed trait does not affect the reproductive potential of the hybrid-derived weedy relative compared to the non-transgenic hybrid-derived population	Propagule production and/or competitive ability of the hybrid-derived weedy relative compared to the non-transgenic hybrid-derived population Increased abundance and distribution of the hybrid-derived weedy relative
Increased fitness of the hybrid-derived weedy relative confers a competitive advantage over a valued species compared to the non-transgenic hybrid-derived population	Increased fitness of the hybrid-derived weedy relative does not affect a valued species compared to the non-transgenic hybrid-derived population	Level of competition between the valued species and the non-transgenic hybrid
The population viability of a valued species is decreased in the local habitat		

Note:

1. Since vertical gene flow is an exposure pathway, not an impact, the only assessment endpoint of gene flow is the occurrence of a transgene in the recipient population. This is reflected in the table with the demarcation of the gene flow exposure pathway.

It is important to note that examples of information elements in this table are intended to illustrate the types of information that can be used in evaluating a risk hypothesis, i.e. to determine whether particular pathway steps are likely to occur. However, for any step there might be other information that would be relevant. Rationales for how such information elements may be used to evaluate the risk hypotheses include:

- The presence of a sexually-compatible weedy relative in the receiving environment provides information as to whether it exists near transgenic plants;
- Flowering time of the transgenic plant and sexually- compatible weedy relative within the receiving environment provides information on overlapping phenology;
- Known hybridisation between the comparator and the sexually-compatible weedy relative and occurrence of natural hybridisation between the transgenic plant and weedy relative (e.g. indicated by a phenotypic or genotypic marker) provide information regarding the probability of fertile hybrid formation;
- The nature of the trait and phenotype of the transgenic plant inform identification of potential adverse effects;
- Presence of the transgene and fitness-associated trait(s) in the hybrid-derived weedy population provide information on the potential degree of phenotypic change affecting population fitness of the hybrid-derived weedy relative;
- Propagule production and/or competitive ability of the hybrid-derived weedy relative compared to the non-transgenic hybrid-derived population provides information on the impact of the transgene on reproductive potential of the hybrid-derived weedy relative. Increased abundance and distribution of the hybrid-derived weedy relative provides information on the impact of the transgene on the fitness of the sexually- compatible weedy relative;
- Level of competition between the valued species and the hybrid-derived weedy species containing transgene compared to non-transgenic hybrid-derived provides information on the potential for the non-transgenic hybrid hybrid-derived population provides information on the relative fitness of the hybrid containing transgene.

References

Arnold, M.L. (1997), *Natural Hybridization and Evolution*, Oxford Series in Ecology and Evolution, Oxford University Press, New York.

Ellstrand, N.C. et al., (2013), "Introgression of Crop Alleles into Wild or Weedy Populations" *Annual Review of Ecology, Evolution, and Systematics.* 2013. Vol. 44, pp. 325-345, https://www.annualreviews.org/doi/10.1146/annurev-ecolsys-110512-135840.

Harrison, R.G. (1990), "Hybrid zones: Windows on evolutionary process", *Oxford Surveys in Evolutionary Biology*, Vol. 7, pp. 69–128, https://www.cabdirect.org/cabdirect/abstract/19910191041.

Kwit, C. et al., (2011), "Transgene introgression in crop relatives: molecular evidence and mitigation strategies", *Trends in Biotechnology*, Vol. 29, pp. 284-293 https://doi.org/10.1016/j.tibtech.2011.02.003.

Annex C. Organisms (Animals)

A consideration for the environmental risk/safety assessment of a transgenic plant is the evaluation of the potential for it to have adverse effects, relative to the comparator, on: (1) animals in the environment and on their role in ecological functions, including food webs; and/or (2) human/animal health due to non-dietary exposure.

Concepts and terms

Plants interact with many other organisms in the environment in a variety of ways. This consideration focuses on a subset of organisms – animals (invertebrates and vertebrates) – that a transgenic plant may interact with, particularly those that may have a role in ecological functions (including food webs), in managed ecosystems (i.e. beneficial organisms in agriculture) or in the wider environment (unmanaged ecosystems). The interactions generally considered include feeding on transgenic plant material by non-domesticated animals, but also encompass non-dietary exposure to animals (including humans). To avoid confusion, the humans are specifically indicated when they are the subject of consideration. Other types of interactions of transgenic plants with organisms are dealt with in other annexes: Annex A (Invasiveness and Weediness, e.g. plant competition); Annex B (Vertical Gene flow, i.e. to other plants); Annex D (Soil Functions, e.g. micro-organisms); Annex E (Plant Health, e.g. pests and pathogens); Annex F (Crop Management Practices, e.g. other organisms in crop fields); and Annex G (Biodiversity, e.g. protected species).

Ecological functions are those functions that an organism, population or community contributes to in the ecosystem in which it resides. Ecological functions include processes, such as pollination, decomposition and nutrient cycling, and the role of organisms as a food source in food webs. Ecological functions become ecosystem services when humans benefit from these functions (Sodhi and Erlich, 2010). Examples of ecosystem services for humankind are pest control by natural enemies (i.e. biological pest control), pollination (e.g. increased fruit set and yield from honeybee activity), soil fertility (e.g. supported by invertebrate detritivores such as springtails) and recreation (e.g. bird watching).

Feeding is the consumption or uptake: of growing or dead plant material by organisms (e.g. by herbivores, pollen consumers and decomposers); or of organisms that have directly fed on plant material (e.g. by parasitoids, scavengers or predators). Dietary considerations associated with the use of transgenic plants as food by humans or feed for domesticated animals, including livestock, are beyond the scope of this document. They are more appropriately addressed in the work programmes of the OECD Working Party for the Safety of Novel Foods and Feeds (WP-SNFF) and the Codex Alimentarius Commission. However, incidental feeding by animals, including non-domesticated animals, on plants or plant parts never intended for use as human food or feed for domesticated animals (e.g. potato meant for industrial starch production, plantation trees, ornamentals) would be relevant for this consideration.

Non-dietary exposure to animals (including humans) may result from any route other than direct feeding, such as dermal contact with the transgenic plant or plant parts or an inhalation exposure to pollen or plant dusts (e.g. from harvesting or processing). Non-dietary exposure also includes interactions via plant structures (e.g. trichomes of stinging nettles) or repellents that prevent herbivore attack (e.g. Agarwal and Rastogi, 2008). Non-dietary exposure may be relevant for human and animal health.

Problem formulation

For this consideration, below is a simple example that illustrates the approach for planning an environmental risk/safety assessment. It includes a discussion of assessment endpoints, potential adverse effects, and a linear pathway to harm with corresponding risk hypotheses and information elements to illustrate the approach. As previously indicated in the document (section 1.2.6), the process is often more complex.

(a) Determination of assessment endpoints

A transgenic plant may have impacts on individual organisms (animals) with which it interacts. These impacts may then affect populations of the species and subsequently ecological functions. Therefore, potential adverse effects at the level of individuals are usually addressed first via tiered testing (Romeis et al., 2011). Impacts at population level on ecological functions and the food web are only expected to arise if the abundance, reproductive biology or behaviour of an organism is affected.

Two examples of assessment endpoints for organisms (animals) are: (1) the quality of the ecological functions of non-domesticated animals (e.g. in pollination; as food source; as beneficial insects, such as ladybird beetles); and (2) human health (i.e. allergic/toxic responses) as a result of non-dietary exposure.

(b) Identification of potential adverse effects on the assessment endpoints

The identification of potential adverse effects of a transgenic plant to an animal considers characteristics of the transgenic plant linked to the genetic modification (e.g. trait, phenotype), and the potential receiving environments.

The potential adverse effects of a transgenic plant on an animal may derive directly from the trait in the transgenic plant. This may include novel proteins (e.g. Cry proteins from *Bacillus thuringiensis*) or double-stranded RNA (dsRNA) that are intended to control a target pest, as well as compounds that repel pest species. Such newly expressed gene products or compounds may also affect animals other than the target pests in terms of survival (i.e. lethal effect), growth, development, reproduction (i.e. sub-lethal effects), behaviour or health (e.g. Romeis et al., 2011).

Potential adverse effects of a transgenic plant on an animal may also derive from intentional or unintentional changes to the plant's composition (e.g. change in levels of endogenous toxicants[1]), morphology (e.g. trichomes) or other characteristics (e.g. changes to response mechanisms of the transgenic plant that are consequences of changes to metabolic pathways). If there is a plausible basis for such changes, then a compositional analysis and phenotypic characterisation can be useful in highlighting differences between the transgenic plant and the comparator, and analysis of differences may suggest a pathway to harm that warrants further consideration.

Consideration of the altered characteristics of the transgenic plant aids in identifying potential adverse effects on assessment endpoints associated with animals. Depending on the changed characteristics that warrant further consideration, potential adverse effects according to the assessment endpoints may include: (1) reduced quality of ecological functions of an animal (e.g. pollination); and (2) increased allergic/toxic responses in humans from non-dietary exposure.

(c) Identification of plausible pathways to harm, formulation of risk hypotheses, and identification of information elements relevant to evaluating the risk hypotheses

In this section, two plausible pathways to harm are postulated. For each step of the postulated pathways to harm, a corresponding risk hypothesis is formulated that will enable the risk assessor to determine whether the pathway is likely to occur. Once it is shown that any part of the pathway is highly unlikely, one does not need to continue evaluating the subsequent steps in the pathway and can conclude that the specific pathway to harm is unlikely to occur. In addition, examples of information elements that can be used to evaluate the risk hypotheses are given along with their rationales.

Postulated pathway leading to reduced quality of ecological function of an animal

Animals present in an agricultural field or in field margins and surroundings can interact with the transgenic plant through direct exposure (e.g. a herbivore feeds directly on the transgenic plant or an animal is affected by repellents produced by the transgenic plant) or indirect exposure (e.g. a parasitoid or predator feeds on herbivores that have fed on the transgenic plant). If the transgenic plant has a changed phenotype (see (b)) that could change the abundance of an animal (e.g. an insect pollinator) this could lead to reduced quality of ecological functions (e.g. reduced pollination of plants in the field and/or field margins that depend on pollination for reproduction).

One example of a postulated pathway to harm for this adverse effect is shown in the first column of Table A C.1. Risk hypotheses for each step of the pathway are formulated in the second column and the third column provides examples of information elements for evaluating the hypotheses.

Table A C.1. Postulated pathway leading to reduced quality of an ecological function, corresponding risk hypotheses, and relevant information elements

Pathway steps	Risk hypotheses	Examples of information elements
In relation to the comparator, the transgenic plant produces a novel gene product in pollen	In relation to the comparator, the transgenic plant does not produce a novel gene product in pollen	Expression of a novel gene in pollen
The pollinator ingests the novel gene product in pollen	The pollinator does not ingest the novel gene product in pollen	Level of expression of the novel gene product; Level of exposure of pollinator to the novel gene product during flowering
The novel gene product has toxic properties for the pollinator when ingested	The novel gene product has no toxic properties for the pollinator when ingested	Nature of the trait of the transgenic plant; Survival, behaviour and reproduction of the pollinator exposed to pollen and/or novel gene product of the transgenic plant
The abundance of the pollinator in the environment is adversely reduced	The abundance of the pollinator in the environment is not adversely reduced	Information on abundance of the pollinator; Other factors influencing the abundance of the pollinator
Pollination is adversely reduced	Pollination is not adversely reduced	Information on reduction in pollination (e.g. seed production, abundance of plants that depend on the pollinator)
Quality of the ecological function of pollination is reduced		

It is important to note that examples of information elements in this table are intended to illustrate the types of information that can be used in evaluating a risk hypothesis, i.e. to determine whether particular pathway steps are likely to occur. However, for any step there might be other information that would be relevant. Rationales for how such information elements may be used to evaluate the risk hypotheses include:

- Expression of the novel gene in the pollen of the transgenic plant as this provides information on whether there is a relevant pathway for exposure to a pollinator;
- The expression level of the novel gene product in pollen and the level of exposure of pollinator to the novel gene product as this provides information on the magnitude of exposure and on whether the pollinator is exposed to sufficient amounts of the protein to adversely affect it;
- The nature of the introduced trait and the phenotype of the transgenic plant informs identification of potential adverse effects (e.g. any insecticidal properties of the novel gene product that could result in direct toxicity to the pollinator). Survival, behaviour and reproduction of the pollinator exposed to the novel gene product as this provides information on the potential adverse effects of that novel gene product to the pollinator. Such data are typically generated using a tiered testing approach in the laboratory;
- Information on the abundance of the pollinator and other factors influencing its abundance (e.g. climatological conditions, current insecticide use and presence of food sources other than the transgenic plant) as this provides information on the impact of the transgenic plant on the pollinator;
- Reduction in pollination (e.g. seed production in plants that depend on the pollinator) as this provides information on whether and by how much pollination capacity is reduced.

Postulated pathway leading to increased allergic/toxic responses in humans from non-dietary exposure

Humans can come into contact with the transgenic plant through non-dietary exposure by way of inhalation of pollen, dermal exposure to plant material during cultivation, or dust during harvest and processing. Such interactions may result in allergic/toxic responses (e.g. allergic symptoms to grain dust exposure (Manfreda et al., 1986)). If the transgenic plant has a changed contact toxicity or allergenicity profile, respiratory or dermal contact may lead to an increased level of dermal and inhalation reactions relative to the comparator.

One example of a postulated pathway to harm for this adverse effect is shown in the first column of Table A C.2. Risk hypotheses for each step of the pathway are formulated in the second column and the third column provides examples of information elements for evaluating the hypotheses.

Table A C.2. Postulated pathway leading to increased allergic/toxic responses in humans from non-dietary exposure, corresponding risk hypotheses, and relevant information elements

Pathway steps	Risk hypotheses	Examples of information elements
In relation to the comparator, the transgenic plant produces a novel protein	The transgenic plant does not produce a novel protein	Production of novel protein
Humans are exposed to the novel protein via non-dietary means	Humans are not exposed to the novel protein via non-dietary means	Routes of non-dietary exposure; Level and pattern of expression of the novel protein in transgenic plant
The novel protein has a human toxicity or allergenicity potential	The novel protein does not have a human toxicity or allergenicity potential	Similarity of novel protein to known human allergens/toxins; Available results of toxicity studies
Toxicity or allergenicity is increased	Toxicity or allergenicity is not increased	Experience with handling the transgenic plant; Data on allergenicity
Toxic or allergic responses in humans are increased due to non-dietary exposure		

It is important to note that examples of information elements in this table are intended to illustrate the types of information that can be used in evaluating a risk hypothesis, i.e. to determine whether particular pathway steps are likely to occur. However, for any step there might be other information that could be relevant. Rationales for how such information elements may be used to evaluate the risk hypotheses include:

- Production of novel proteins in the transgenic plant (e.g. from expression of the introduced trait or from novel open reading frames created by insertion of the DNA sequences) as this provides information on whether there are novel proteins expressed in the transgenic plant compared to the comparator;
- Routes of non-dietary exposure of humans to the transgenic plant or plant parts as this provides information on the interaction between the transgenic plant and humans;
- Similarity of the novel protein(s) to known human allergens or toxins (e.g. via bioinformatic analysis) as this provides information on whether the transgenic plant has a human toxicity or allergenicity potential;
- Experience with the handling of the transgenic plant, including any reports of toxic or allergenic effects, and information from allergenicity assessment (e.g. sera screening) and toxicity laboratory studies with animals as this provides information on whether there are increased allergenic or toxic effects.

References

Argarwal V.M. and N. Rastogi (2008), “Role of floral repellents in the regulation of flower visits of extrafloral nectary-visiting ants in an Indian crop plant”. *Ecological Entomology*, Vol. 33, pp. 59-65, https://doi.org/10.1111/j.1365-2311.2007.00938.x.

Manfreda, J. et al. (1986), “Acute symptoms following exposure to grain dust in farming”, *Environmental Health Perspectives*, Vol. 66, pp. 73-80, https://doi.org/10.1289/ehp.866673.

Romeis, J. et al. (2011), “Recommendations for the design of laboratory studies on non-target arthropods for risk assessment of genetically engineered plants”, *Transgenic Research*, Vol. 20, pp. 1-22, https://doi.org/10.1007/s11248-010-9446-x.

Sodhi, N.S. and P.R. Ehrlich (2010), *Conservation Biology for All*, Oxford University Press, https://doi.org/10.1093/acprof:oso/9780199554232.001.0001.

Note

[1] The Consensus documents on plant composition issued by the OECD WP-SNFF contain information on endogenous toxicants, allergens and anti-nutrients, https://www.oecd.org/chemicalsafety/biotrack/consensus-document-for-work-on-safety-novel-and-foods-feeds-plants.htm.

Annex D. Soil Functions

A consideration for the environmental risk/safety assessment of a transgenic plant is the evaluation of the potential for the plant to have adverse effects on soil microbial communities responsible for soil processes and their soil functions, relative to the comparator.

Concepts and terms

Soil functions, such as soil quality, primarily depend on biotic factors including soil flora and fauna and their abundance and composition. Fauna ranging from micro- through meso- and macro- to megafauna and the associated soil processes are critical to maintain soil quality. This environmental consideration focuses on the interactions of transgenic plants with soil micro-organisms, and the potential effects on soil quality, soil biogeochemical cycling, or other microbe-mediated soil processes when soil microbes are adversely affected. Considerations in Annex C (Organisms (Animals)) can be applied to higher soil fauna, such as arthropods and nematodes.

Soil quality has many definitions depending on the context, national legal frameworks, and the soil science community. Soil quality reflects, *inter alia*, the potential of the soil to sustain plant growth and the above-ground ecosystem by providing nutrients and minerals, by providing microbial factors involved in plant health (e.g. absence of pathogens or presence of antagonists of plant pathogens), and by safeguarding microbial functional diversity. The requirements for soil quality may be different in different ecosystems.

Biogeochemical cycling processes have important ecological functions for the maintenance of soil quality. Examples of such processes are mineralisation, nitrification, carbon (C)-cycling, nitrogen (N)-fixation, soil respiration, decomposition of organic matter, and humification. It is widely recognised that microbial communities in soils, which are known to be important for biogeochemical cycling processes, vary considerably both temporally and spatially. Biogeochemical cycling processes are relatively robust to changes in soil microbial community structure (abundance and diversity of species) due to redundancy in microbial community function.

Problem formulation

For this consideration, below is a simple example that illustrates the approach for planning an environmental risk/safety assessment. It includes a discussion of assessment endpoints, potential adverse effects, and a linear pathway to harm with corresponding risk hypotheses and information elements to illustrate the approach. As previously indicated in the document (section 1.2.6), the process is often more complex.

(a) Determination of assessment endpoints

Soil microbial communities are very complex, often characterised by high microbial diversity (Tiedje et al., 1999; Roesch-Luiz et al., 2007; Fierer and Lennon, 2011), and in constant flux in response to several factors (Leitner, Aaron and Jodi, 2021). The occurrence and abundance of soil micro-organisms are affected by 1) soil characteristics like organic matter content, nutrient content, and moisture capacity; 2) typical physico-chemical factors such as temperature, pH, redox potential and physical soil structure;

and 3) influences caused by human activities like crop rotation, soil management practices and chemical control methods. Soils are heterogeneous and significant variation in microbial populations is expected in soil, including in agricultural fields.

Plants have impacts on soil micro-organisms with which they interact. These interactions may then affect soil microbial communities. Such a change could also affect soil processes underlying soil quality.

One example of an assessment endpoint for soil functions is the quality of soil.

(b) Identification of potential adverse effects on the assessment endpoints

The identification of potential adverse effects of the transgenic plant on soil functions considers the characteristics of the transgenic plant linked to the genetic modification, including the novel gene product(s), and the potential receiving environments.

A transgenic plant may express a gene product or produce a new metabolite based on the expression of a gene product. This may cause adverse effects on soil functions in different ways, for example by affecting the diversity of microbial species and/or soil microbial communities, or by affecting biogeochemical cycling processes. Plant growth and health may be compromised by impaired soil functions.

Potential impacts of a transgenic plant on soil quality – and more generally on soil functions – via crop management practices are taken into account in Annex F (Crop Management Practices).

Other impacts of a transgenic plant on soil quality – and more generally on soil functions – may occur due to a potential plant to micro-organisms gene transfer (for example in the case of antibiotic resistance genes), whose corresponding risk assessment may be required in some jurisdictions but will not be elaborated further in this document.

Consideration of the mechanism of action of the newly introduced trait and the characteristics of the transgenic plant relative to its comparator aids in identifying potential adverse effects on soil functions. An example of an adverse effect on the environment according to the assessment endpoint identified above is reduction of soil quality.

(c) Identification of plausible pathways to harm, formulation of risk hypotheses, and identification of information elements relevant to evaluating the risk hypotheses

In this section, a plausible pathway to harm is postulated. For each step of the postulated pathway to harm, a corresponding risk hypothesis is formulated that will enable the risk assessor to determine whether the pathway is likely to occur. Once it is shown that any part of the pathway is highly unlikely to occur, one does not need to continue evaluating the subsequent steps in the pathway and can conclude that the specific pathway to harm is unlikely to occur. In addition, examples of information elements that can be used to evaluate the risk hypotheses are given along with their rationales.

Postulated pathway leading to a reduction in soil quality

Soil micro-organisms and/or microbial communities can be exposed to a gene product or new metabolite produced by a transgenic plant via root exudation or by leaching from plant parts that are shed onto or into the soil. If the gene product or new metabolite has the capability to directly affect certain soil micro-organisms and/or microbial communities (e.g. a transgenic plant with a disease resistance trait), this may lead to changes in biogeochemical processes and in the end could lead to an altered soil quality.

One example of a postulated pathway to harm for this adverse effect is shown in the first column of Table A D.1. Risk hypotheses for each step of the pathway are formulated in the second column and the third column provides examples of information elements for evaluating the hypotheses.

Table A D.1. Postulated pathway leading to a reduction in soil quality, corresponding risk hypotheses, and relevant information elements

Pathway steps	Risk hypotheses	Examples of information elements
In relation to the comparator, the transgenic plant produces a new substance that has antimicrobial properties against certain soil micro-organisms	In relation to the comparator, the transgenic plant does not produce a new substance that has antimicrobial properties against certain soil micro-organisms	The intended function of the DNA sequences of the transgene in the transgenic plants are to produce an antimicrobial protein or metabolite
The new substance is released into the soil	The new substance is not released into the soil	Level and pattern of expression of the novel substance in the transgenic plant Soil stability and fate of novel substance in the soil
Abundance and diversity of the soil micro-organisms affected by the new substance are reduced	Abundance and diversity of the soil micro-organisms affected by the new substance are not reduced	Population dynamics of soil micro-organisms
Soil quality due to microbial activity in the soil is affected, e.g. reduced	Soil quality due to soil micro-organisms is not affected	Effects on processes such as for example ammonification
Key soil processes due to activities of beneficial soil micro-organisms are persistently disrupted	Soil processes due to activities of beneficial soil micro-organisms are not persistently disrupted	Role of micro-organisms in disrupted biogeochemical processes Functional redundancy among soil micro-organisms Length of time to soil processes recovery
Soil property is persistently reduced by the transgene in the cultivation of the transgenic plant		

It is important to note that examples of information elements in this table are intended to illustrate the types of information that can be used in evaluating a risk hypothesis, i.e. to determine whether particular pathway steps are likely to occur. However, for any step there might be other information that could be relevant. Rationales for how such information elements may be used to evaluate the risk hypotheses include:

- The DNA sequences introduced into the transgenic plant, any novel proteins or metabolites produced in the transgenic plant, and the antimicrobial properties of the novel gene product/metabolite, which inform the potential for the transgenic plant to produce novel anti-microbial substances;
- The level and pattern of expression of the novel substance in the transgenic plant and its stability and fate in the soil (e.g. rapidly degraded or persistent), which informs the level and duration of exposure of the soil micro-organisms to the novel substance;
- The changes in microbial activity in the soil related to soil processes (e.g. ammonification), which provide insight on the level of impact of these changes to the soil quality;
- The role of the micro-organisms in disrupted biogeochemical processes including supporting plant growth, the functional redundancy among soil micro-organisms and the length of time of soil processes recovery, which informs the likelihood and magnitude of the impact on key biogeochemical processes due to persistent disruption of activities of beneficial soil micro-organisms.

References

Fierer, N. and J.T. Lennon (2011), "The generation and maintenance of diversity in microbial communities", *American Journal of Botany*, Vol. 98, (3), pp. 439-448, https://doi.org/10.3732/ajb.1000498.

Roesch, L.F.W. et al. (2007), "Pyrosequencing enumerates and contrasts soil microbial diversity", *ISME Journal*, Vol. 1, (4), pp. 283-290, https://doi.org/10.1038/ismej.2007.53.

Tiedje, J.M. et al. (1999), "Opening the black box of soil microbial diversity", *Applied Soil Ecology*, Vol. 13, (2), pp. 109-122, https://doi.org/10.1016/S0929-1393(99)00026-8.

Leitner, Z.R., L.M. Aaron and D.H. Jodi (2021), "Temporal fluctuations of microbial communities within the crop growing season", *Geoderma*, Vol. 391, 114951, https://doi.org/10.1016/j.geoderma.2021.114951.

Annex E. Plant Health

A consideration for the environmental risk/safety assessment of a transgenic plant is the evaluation of its potential to have adverse effects on plants in the environment. Relative to the comparator, the transgenic plant may have an unintended increased susceptibility to pests, which may impact plant health in agroecosystems.

Concepts and terms

For the purpose of this annex, we refer to managed and areas adjacent to the cultivated field as the agroecosystem. Therefore, the health of cultivated and other valued plants in the agroecosystem is a consideration in the environmental risk/safety assessment of a transgenic plant.

Plant health refers to a plant's capacity to express its full genetic potential as a valued plant in an agroecosystem. Ideal expression of plant health is the result of optimally exhibited desirable phenotypic traits, such as growth and development or vegetative or reproductive yield.

Pest includes any species, strain, or biotype of plant, animal (e.g. insect), or pathogenic agent (e.g. microbe) injurious to plants or plant products; IPPC, 2021). Pests which are a plant (e.g. weed) or vertebrate animal (e.g. rodent) are beyond the scope of this section.

Host plant is a plant that may harbour a specific pest, depending on that plant's susceptibility.

Susceptibility refers to a plant's inability to restrict the growth and development of a given pest.

Problem formulation

For this consideration, below is a simple example that illustrates the approach for planning an environmental risk/safety assessment. It includes a discussion of assessment endpoints, potential adverse effects, and a linear pathway to harm with corresponding risk hypotheses and information elements to illustrate the approach. As previously indicated in the document (section 1.2.6), the process is often more complex.

(a) Determination of assessment endpoints

Plants have an innate ability to resist pests. Cultivated plants tend to have a reduced ability to resist pests relative to wild plants (Whitehead et al., 2017). Plant breeders have successfully developed pest-resistant varieties through conventional breeding and through transgenic approaches. Selection for traits other than pest-resistance, such as low-lignin to increase forage quality, may increase pest susceptibility in the cultivated plant. Depending on the plant, trait, and environment, a transgenic plant may have an unintended increased susceptibility to certain pests, which may impact plant health in the agroecosystem relative to the comparator.

Operational protection goals derived from the general protection goals (e.g. to protect plant health) may be used as an intermediate step to facilitate the selection of assessment endpoints. An example of an operational protection goal could be to minimise or prevent injury to cultivated and other valued plants in the agroecosystem by pests associated with the transgenic plant. An example of a relevant assessment endpoint for the operational protection goal is the health of cultivated and other valued plants in the agroecosystem.

(b) Identification of potential adverse effects on the assessment endpoints

Plant health may be impacted by the introduction of a cultivated plant to the agroecosystem. Whether adverse effects occur depends on the receiving environment and the phenotypic change. Although the presence of a transgene in a plant does not inherently increase the likelihood of adverse effects on the environment, the phenotypic change derived from the transgene may alter the transgenic plant's impacts relative to the comparator. Phenotypic change associated with the transgene may include novel gene products, modified biochemical components (e.g. polyphenols), or alterations in plant protective architecture (e.g. lignin). These changes may affect the transgenic plant's interaction with pests and lead to adverse effects on plant health in the agroecosystem relative to the comparator.

A phenotypic change, whether derived from conventional breeding or a transgene, may increase the plant's susceptibility to pests and even enhance its capacity to harbour pests. Consequently, if the pest load increases in the agroecosystem, pests could spread and adversely affect plants, either in the same growing season as the transgenic plant or in subsequent seasons, even in the absence of the transgenic plant. For example, pests such as cereal rusts, nematodes, and soil-borne pathogens may survive and spread via alternative host plants or volunteers in the agroecosystem (Zeng and Luo, 2006; Baley et al., 2008; CABI and USDA, 2018).

Thus, depending on the changes in phenotype of the transgenic plant relative to a comparator, a potential adverse effect on plant health according to the assessment endpoint identified may include the decreased viability of plants in the ecosystem due to increased susceptibility of the transgenic plant to certain pests.

(c) Identification of plausible pathways to harm, formulation of risk hypotheses, and identification of information elements relevant to evaluating the risk hypotheses

In this section, a plausible pathway to harm is postulated. For each step of a postulated pathway to harm, a corresponding risk hypothesis is formulated that will enable the risk assessor to determine whether the pathway is likely to occur. Once it is shown that any step of the pathway is highly unlikely to occur, one does not need to continue evaluating the subsequent steps in the pathway and can conclude that the specific pathway to harm is unlikely to occur. In addition, examples of information elements that can be used to evaluate the risk hypotheses are given along with their rationales.

Postulated pathway leading to decreased viability of a cultivated and/or other valued plant in an agroecosystem

Pests occurring in an agroecosystem may use the transgenic plant, as with any plant, as a host, with or without associated impacts. If the transgenic plant has a changed phenotype (e.g. biochemical composition or plant protective architecture), this change could lead to a new or modified niche for pests, which could lead to an increase in pest abundance and, ultimately, injury to cultivated and/or other valued plants in the agroecosystem in the same and/or subsequent growing season(s) as the transgenic plant.

One example of a postulated pathway to harm for this adverse effect is shown in the first column of Table A E.1. Risk hypotheses for each step of the pathway are formulated in the second column and the third column provides examples of information elements for evaluating the hypotheses.

Table A E.1. Postulated pathway leading to decreased viability of a cultivated and/or other valued plant in an agroecosystem due to increased susceptibility of the transgenic plant, corresponding risk hypotheses, and relevant information elements

Pathway steps	Risk hypotheses	Examples of information elements
The modified trait in the transgenic plant alters the plant's chemical or structural defence mechanisms	There is no alteration in the plant's chemical or structural defence mechanisms	Expression and nature of the introduced gene product and its function in the transgenic plant
The transgenic plant exhibits pest susceptibility directly in terms of increased disease symptoms or damage and/or indirectly through increased pest numbers (harbouring pests) on the transgenic plant, relative to the comparator	There are no increased disease symptoms or insect damage and/or pest numbers on the transgenic plant relative to the comparator	Changes in pest populations and incidence of pest-damage to the transgenic plant and comparator
The transgenic plant acts as a greater source of pests that spread to plants in the agroecosystem in the same or subsequent growing season(s) relative to the comparator	The transgenic plant does not act as a greater source of pests for plants in the agroecosystem in the same or subsequent growing season(s) relative to the comparator	Plants present in the agroecosystem; known susceptibility of these plants to the pests; and changes in pest populations on plants in the agroecosystem in the same or subsequent growing season(s)
There is an increase in injury to plants in the agroecosystem by pests in the same or subsequent growing season(s) relative to the comparator	There is no increase in injury to plants in the agroecosystem by pests in the same or subsequent growing season(s) relative to the comparator	Injury/damage incidence to plants in the agroecosystem in the same or subsequent growing season(s) relative to the comparator
The viability of a cultivated and/or other valued plant in the agroecosystem is reduced		

It is important to note that examples of information elements in this table are intended to illustrate the types of information that can be used in evaluating a risk hypothesis, i.e. to determine whether particular pathway steps are likely to occur. However, for any step there might be other information that could be relevant. Rationales for how such information elements may be used to evaluate the risk hypotheses include:

- Expression and nature of the introduced gene product and its function in the transgenic plant provide information as to whether a phenotypic change may affect the transgenic plant's defence mechanisms;
- Changes in pest populations and incidence of pest-damage to the transgenic plant and comparator provide information as to whether there is a difference in pest susceptibility between the transgenic plant and the comparator;
- Plants present in the agroecosystem; known susceptibility of these plants to the pests; and changes in pest populations on plants in the agroecosystem in the same or subsequent growing season(s) provide information on whether the transgenic plant may be a source of pests that spread to other plants;
- Injury/damage incidence to plants in the agroecosystem in the same or subsequent growing season(s) relative to the comparator provides information on whether there has been an increase in pest-related damage to plants in the agroecosystem relative to the comparator.

References

Baley, G.K. et al. (2008), "Influence of glyphosate, crop volunteer, and root pathogens on glyphosate-resistant wheat under controlled environmental conditions", *Pest Management Science*, Vol. 65, pp. 288-299, https://doi.org/10.1002/ps.1687.

CAB International and USDA (2018), "Plant Parasitic Nematodes", in R. Sikora et al. (eds), *Subtropical and Tropical Agriculture*, Boston, MA: CABI, Plant parasitic nematodes in subtropical and tropical agriculture | CABI Books (cabidigitallibrary.org).

IPPC (2021), *Pest Status Guide: Understanding the Principal Requirements for Pest Status Determination*, International Plant Protection Convention, Food and Agriculture Organization of the United Nations, https://www.fao.org/documents/card/en/c/cb6103en.

Whitehead, S.R., M.M. Turcotte and K. Poveda (2017), "Domestication impacts on plant–herbivore interactions: a meta-analysis", *The Royal Society Publishing*. Vol. 372, https://doi.org/10.1098/rstb.2016.0034.

Zeng, S.M. and Y. Luo (2006), "Long-distance spread and interregional epidemics of wheat stripe rust in China", *American Phytopathological Society: Plant Disease*, Vol. 90, (8), pp. 980-988, https://doi.org/10.1094/PD-90-0980.

Annex F. Crop Management Practices

A consideration for the environmental risk/safety assessment of a transgenic plant is the evaluation of the potential for changes in crop management practices associated with its cultivation relative to those associated with the cultivation of the comparator, and if such changes could have adverse effects on the environment.

Concepts and terms

Crop management practices are agricultural practices used to increase the quality and yield of crops. Examples of crop management practices include: soil tillage; crop rotation; irrigation; fertilisation; as well as mechanical, biological, cultural, and chemical methods for managing weeds (including volunteers) and pests (including insects and diseases). The combination, timing, and sequence of the crop management practices used by farmers varies based on factors such as the crop species and its growth stage, the soil, climatic and weather conditions, pest pressure, and socio-economic factors.

Tillage is a crop management practice involving the preparation of soil by mechanical disturbance, such as digging, stirring, and overturning. Tillage is widely used to incorporate crop residues and manure into the soil, limit the growth of weeds during the intercropping period, prepare a seedbed, and control weeds in crop fields. Conventional tillage leaves the surface of the field relatively bare and susceptible to wind and water erosion. To avoid this risk, farmers have increasingly used reduced-till or no-till practices that are collectively referred to as conservation tillage practices because they protect the soil surface. A reduced-till system retains more crop residue cover than conventional tillage whereas a no-till system leaves the crop residue undisturbed from harvest through planting. Conservation tillage practices have been supported by agronomic developments such as herbicides for weed control, herbicide-tolerant crops (transgenic and non-transgenic), and improved farm machinery.

Crop rotation is a crop management practice involving growing different crops in succession in a particular area. Crop rotation can reduce weed and pest pressure, and maintain or restore nutrient balances in the soil.

The term organism is used in this annex for plants, animals, and micro-organisms.

Problem formulation

For this consideration, below are simple examples that illustrate the approach for planning an environmental risk/safety assessment. It includes a discussion of assessment endpoints, potential adverse effects, and a linear pathway to harm with corresponding risk hypotheses and information elements to illustrate the approach. As previously indicated in the document (section 1.2.6), the process is often more complex.

(a) Determination of assessment endpoints

The cultivation of a transgenic plant has the potential to alter crop management practices. However, such alterations are not unique to the cultivation of a transgenic plant. Crop management practices are in constant flux in response to weed and pest pressure, climate and weather, economics, regulations, new crop varieties and technologies, and other forces that impact a farmer's land-use decisions. Potential changes in crop management practices associated with the cultivation of a transgenic plant are only considered in the environmental risk/safety assessment when the potential changes may lead to adverse environmental effects.

Two examples of assessment endpoints for evaluating the potential environmental impact of changes in crop management practices are: (1) the abundance of a valued organism in crop fields or field margins and (2) the quality of soil.

(b) Identification of potential adverse effects on the assessment endpoints

Certain characteristics of a transgenic plant linked to a genetic modification may lead to crop management practices being changed in specific ways. These changes may in turn alter weed and pest populations, for example leading to the evolution of herbicide resistance in weeds or pesticide resistance in target organisms. The adoption of integrated weed and pest management strategies are widely recommended to farmers to manage such issues (e.g. Anderson et al., 2019). Depending on their laws and policies, some countries will also consider the potential for changes in weed and pest populations prior to the release of a transgenic plant.

The identification of potential adverse effects on the environment resulting from changes in crop management practices should be informed by knowledge of the range of existing crop management practices used for the comparator and whether the characteristics of the transgenic plant might affect these practices. Agronomic studies can be useful in highlighting the differences in crop management practices between the transgenic plant and the comparator. However, caution should be exercised in interpreting such differences given that crop management practices are influenced by a wide range of factors (mentioned above) and ultimately determined by the farmer. Consequently, potential adverse effects from changes in crop management practices cannot always be predicted solely from the new characteristics of the transgenic plant because crop management practices involve many factors not directly related to the transgenic plant.

Nevertheless, consideration of the characteristics of the transgenic plant relative to the comparator aids in identifying potential adverse effects associated with changes in crop management practices. Thus, depending on the changes in characteristics of the transgenic plant in relation to the comparator, two examples of potential adverse effects on the environment according to the assessment endpoints identified above may include: (1) reduced abundance of a valued organism and (2) reduced quality of soil.

(c) Identification of plausible pathways to harm, formulation of risk hypotheses, and identification of information elements relevant to evaluating the risk hypotheses

In this section, two plausible pathways to harm are postulated. For each step of the postulated pathways to harm, a corresponding risk hypothesis is formulated that will enable the risk assessor to determine whether the pathways are likely to occur. Once it is shown that any step of the pathway is highly unlikely to occur, one does not need to continue evaluating the subsequent steps in the pathway and can conclude that the specific pathway to harm is unlikely to occur. In addition, examples of information elements that can be used to evaluate the risk hypotheses are given along with their rationales.

Postulated pathway leading to reduced abundance of a valued organism

Cultivation of a transgenic plant with a trait (e.g. herbicide tolerance) that leads to changes in weed management practices may reduce the number of wild plants present in crop fields and field margins or induce population shifts in those wild plants. If a valued organism depends on those wild plants occurring in the field or in field margins for food or habitat, such a change in crop management practices may result in reduced abundance of a valued organism.

One example of a postulated pathway to harm for this adverse effect is shown in the first column of Table A F.1. Risk hypotheses for each step of the pathway are formulated in the second column and the third column provides examples of information elements for evaluating the hypotheses.

Table A F.1. Postulated pathway leading to reduced abundance of a valued organism, corresponding risk hypotheses, and relevant information elements

Pathway steps	Risk hypotheses	Examples of information elements
The introduced trait confers herbicide tolerance	The introduced trait does not confer herbicide tolerance	Identity of the introduced gene; Activity of the new protein; response of the transgenic plant to herbicide applications
The transgenic plant is cultivated within the geographic distribution range of wild plants that are important for a valued organism	The transgenic plant is not cultivated within the geographic distribution range of wild plants that are important for a valued organism	Geographic distribution range of the valued organism; Dietary and habitat needs of the valued organism
There is a change in herbicide regime for the transgenic crop relative to the comparator crop(s)	The transgenic crop is not differently treated relative to the comparator crop(s)	Current herbicide regime used on comparator crop(s); Proposed herbicide regime for the transgenic crop
The abundance of wild plants that support the abundance of valued organisms is reduced by the change in herbicide applied	The abundance of wild plants that support the abundance of valued organisms is not reduced by the change in herbicide applied	Presence of valued organisms in the receiving environment; presence of wild plants that support the abundance of valued organisms in the receiving environment; Effect of the proposed herbicide regime on these plants; Alternate habitats available for valued organisms
The abundance of valued organisms is reduced		

It is important to note that examples of information elements in this table are intended to illustrate the types of information that can be used in evaluating a risk hypothesis, i.e. to determine whether particular pathway steps are likely to occur. However, for any step there might be other information that could be relevant. Rationales for how such information elements may be used to evaluate the risk hypotheses include:

- The identity of the introduced gene, activity of the new protein and response of the transgenic plant when challenged with herbicides provides information on the new herbicide tolerance conferred to the transgenic plant;
- The geographic distribution range of the valued organism and its dietary and habitat needs provides baseline information about its presence and survival needs;
- The current herbicide regime used on the comparator crop(s) and the proposed herbicide regime for the transgenic crop (e.g. which herbicide groups, timing, and frequency of herbicide applications) provides information on any changes in the herbicides applied (e.g. the herbicide the transgenic plant is tolerant to);

- The proposed new herbicide regime and its known or anticipated effect on wild plant populations present in the receiving environment provides information on whether those habitat plants will be susceptible to the new regime; the existence of alternate habitats for the valued organism provides information on whether loss of wild plants in the crop or field margins due to changes in crop management practices may impact the abundance of the valued organism.

Postulated pathway leading to reduced quality of soil

Many crop management practices directly or indirectly impact the quality of soil. The adoption of reduced-till or no-till practices may improve the quality of agricultural soils (Derpsch et al., 2010; Busari et al., 2015). However frequent use of conventional tillage may result in soil compaction or erosion (FAO, 2022). As noted above, use of herbicides for weed control can support minimum tillage agriculture. However frequent application of any herbicide can result in the evolution of herbicide resistant weeds (Owen and Zelaya, 2005). If weed control in a herbicide-tolerant transgenic crop relies on the specific herbicide, then resistant weeds may arise. If this is the case, then crop management practices may revert to tillage for weed control which may lead to a reduction in soil quality.

One example of a postulated pathway to harm for this adverse effect is shown in the first column of Table A F.2. Risk hypotheses for each step of the pathway are formulated in the second column and the third column provides examples of information elements for evaluating the hypotheses.

Table A F.2. Postulated pathway leading to reduced abundance of a valued organism, corresponding risk hypotheses, and relevant information elements

Pathway steps	Risk hypotheses	Examples of information elements
The introduced trait confers tolerance to a specific herbicide	The introduced trait does not confer tolerance to a specific herbicide	Identity of the introduced gene; Activity of the new protein; Response of the transgenic plant to herbicide applications
There is an increased use of the specific herbicide to control weeds in the transgenic crop relative to the comparator crop(s)	There is not an increased use of the specific herbicide to control weeds in the transgenic crop relative to the comparator crop(s)	Current herbicide regime used on comparator crop(s); Proposed herbicide regime for the transgenic crop
The abundance of herbicide-resistant weeds increases due to selection pressure from repeated application of the specific herbicide	The abundance of herbicide-resistant weeds does not increase due to selection pressure from repeated application of the specific herbicide	Identity of target weeds; presence and incidence of herbicide-resistant weed populations; Mitigation measures available to delay development of herbicide resistance in weeds; Mode of action of the specific herbicide
The specific herbicide does not control herbicide-resistant weeds and reliance on conventional tillage for weed control increases	The specific herbicide does control herbicide-resistant weeds and reliance on conventional tillage for weed control does not increase	Available options to control herbicide-resistant weeds
Additional passes of heavy machinery over the field and reductions in crop residues that protect the soil from wind and water erosion result in increased soil compaction, erosion, organic matter loss	There is no increased soil compaction, erosion, organic matter loss due to additional passes of heavy machinery over the field or reductions in crop residues that protect the soil from wind and water erosion	Available measures to maintain soil quality when conventional tillage is used; ability of soil to support crop growth
The quality of soil is reduced		

It is important to note that examples of information elements in this table are intended to illustrate the types of information that can be used in evaluating a risk hypothesis, i.e. to determine whether particular pathway steps are likely to occur. However, for any step there might be other information that could be relevant. Rationales for how such information elements may be used to evaluate the risk hypotheses include:

- The identity of the introduced gene, activity of the new protein and the response of the transgenic plant when challenged with the specific herbicide provide information on the new herbicide tolerance conferred to the transgenic plant;
- The current herbicide regime used on the comparator crop(s) and the proposed herbicide regime for the transgenic crop (e.g. herbicide groups, timing and frequency of herbicide applications) provides baseline information for identifying the changes in herbicide use following the introduction of the transgenic crop, including more frequent use of a particular herbicide or increased amount of acreage treated with a particular herbicide;
- Knowledge of the weed species present in the receiving environment, the ability of these weed species to develop resistance to herbicides in general, the presence of any weed populations already resistant to the specific herbicide, and the mode of action of the specific herbicide provide information about the likelihood of weed species evolving resistance to the specific herbicide;
- Knowledge of the proposed herbicide regime for the transgenic crop and any mitigation measures available to delay the evolution of herbicide-resistance in weed populations also provides information on the likelihood of weed species evolving resistance to the specific herbicide;
- Knowledge of alternatives to the specific herbicide, including other herbicides or non-chemical methods provides information on the level of tillage that may be required to control herbicide resistant weeds and the level of impact on soil;
- The availability of measures to mitigate for the detrimental effects of conventional tillage on soil degradation provides information on the potential extent of reduction in soil quality.

References

Anderson, J.A. et al. (2019), "Genetically engineered crops: Importance of diversified integrated pest management for agricultural sustainability", *Frontiers in Bioengineering and Biotechnology.* Vol. 7, https://doi.org/10.3389/fbioe.2019.00024.

Busari et al. (2015). "Conservation tillage impacts on soil, crop and the environment", *International Soil and Water Conservation Research,* Vol. 3(2), pp. 119-129, https://doi.org/10.1016/j.iswcr.2015.05.002.

Derpsch, R. et al. (2010), "Current status of adoption of no-till farming in the world and some of its main benefits", *International Journal of Agricultural and Biological Engineering*, Vol. 3 (1), pp. 1-25, https://ijabe.org/index.php/ijabe/article/view/223.

FAO (2022), "Three principles of Conservation Agriculture", Food and Agriculture Organization of the United Nations, webpage http://www.fao.org/ag/ca/1a.html (accessed 27 April 2023).

Owen, M.D. and I.A. Zelaya (2005), "Herbicide-resistant crops and weed resistance to herbicides", *Pest Management Science*, Vol. 61, pp. 301-311, https://doi.org/10.1002/ps.1015.

Annex G. Biodiversity (Protected Species and Habitats/Ecosystems)

A consideration for the environmental risk/safety assessment of a transgenic plant is the evaluation of the potential for it to have adverse effects on biodiversity of species and habitats/ecosystems explicitly protected by legislation of a country or a region.

Concepts and terms

Protection of biodiversity at genetic, species, and habitat/ecosystem levels is an overarching issue that is integral to environmental risk/safety assessment.

Addressing the potential environmental impact of the release into the environment of a transgenic plant on biodiversity is approached differently in different jurisdictions. For the purpose of this annex, biodiversity or biological diversity focuses on species and/or geographically defined habitats/ecosystems explicitly protected by legislation of a country or region. Biodiversity is also considered in a number of the other environmental consideration annexes relative to potential impacts of a transgenic plant on the biodiversity of species valued (but not explicitly protected) for a variety of reasons including, but not limited to, their contribution to ecological functions. Elements of a number of the environmental considerations described in the other annexes could also be relevant for evaluating the potential of a transgenic plant to adversely affect explicitly protected species and habitats/ecosystems.

However, the conceptual framework for evaluating the potential for a trangenic plant to adversely affect explicitly protected species differs somewhat from the conceptual framework employed in Annexes A-F. For an explicitly protected species, the conservation of every individual of the protected species is important in maintaining the ability of a species to evolve in response to changing environmental variables and avoid extinction. The need to conserve each individual in the species leads assessors, particularly for those species that are low in numbers and could be harmed by testing (e.g. insects), to equate the number of individuals to the genetic diversity of the species. In general, as the number of individuals in a species dwindles, the species loses diversity. For this reason, the conceptual framework generally employed when considering the genetic diversity of an explicitly protected species is the potential effect of the transgenic plant on each individual of an explicitly protected species. Maintaining the numbers of an explicitly protected species is a measure employed to maintain the genetic diversity of the species and thus helps to avoid species extinction.

Biodiversity, or biological diversity, means the variability among living organisms from all sources including, *inter alia*, terrestrial, marine and other aquatic ecosystems and the ecological complexes of which they are part. This includes the diversity within species, between species and of ecosystems (CBD, 1992).

Explicitly protected species or habitats/ecosystems include, for the purpose of this annex, animal or plant species and/or habitats/ecosystems that are explicitly listed as protected due, for example, to their level of endangerment or threatened status (i.e. national endangered species/critical habitat legislation) or their cultural significance because of the responsibility a country has for the given species (e.g. Commonwealth of Australia, 1999).

Protected habitat/ecosystem means a geographically defined area or areas which is/are designated or regulated and managed to achieve specific conservation objectives (see e.g. CBD, 1992). For the sake of simplicity, "explicitly protected species or habitats/ecosystems" are referred to in the text to follow as "protected species or habitats/ecosystems". In many jurisdictions, species protection is approached in a way that connects the protected species and its habitat/ecosystem, because protected species, in general, are unlikely to survive without the environment that provides the elements necessary for the survival of the species (e.g. the United States' Endangered Species Act 1973 (USA, 1973), Australia's Environment Protection and Biodiversity Conservation Act 1999 (Commonwealth of Australia, 1999), and the European Union's Directive 92/43/EEC 1992 on the Conservation of Natural Habitats and of Wild Fauna and Flora (European Union, 1992)).

Box A G.1. Explicitly protected habitats in the European Union (EU)

In the EU, the Habitats Directive (Council Directive 92/43/EEC on the conservation of natural habitats and of wild fauna and flora) ensures the conservation of a wide range of rare, threatened or endemic animal and plant species. In addition, some 200 rare and characteristic habitat types are also targeted for conservation in their own right, e.g. 'Mesophile grasslands', 'Forests of temperate Europe', 'Sphagnum acid bogs' or 'Thermo-Mediterranean and pre-steppe brush'.

Hence, in the EU habitats that are protected due to their function as living area for protected species and/or due to their own rare or endangered habitat composition. In consequence, both aspects are considered during the risk/safety assessment of a transgenic plant for the release into the environment according to Commission Directive (EU) 2018/350 where protected habitats are referred to as 'protected areas'.

Source: European Union (2018), https://ec.europa.eu/environment/nature/legislation/habitatsdirective/index_en.htm (accessed 12 July 2023).

Box A G.2. Explicitly protected habitats in the United States of America (USA)

During an evaluation of a transgenic plant for experimental testing or commercial use, the protected habitat is considered in the context of designated critical habitat of federally listed threatened and endangered species. Under the USA's Endangered Species Act 1973, critical habitat is the specific areas within the geographic area occupied by the species at the time it was listed, that contain the physical or biological features that are essential to the conservation of endangered and threatened species and that may need special management or protection. Designated critical habitat of listed species is an important consideration as many protected species in general are unlikely to survive without the environment that provides the elements necessary for the survival of the species.

Source: U.S Fish and Wildlife Service (1973), Endangered Species Act.

Genetic diversity represents both an overall protection aim (e.g. in the CBD, 1992) as well as a factor taken into account when evaluating the potential for a transgenic plant to affect a protected species. This is because the ability to adapt to changing environmental conditions is vital for the survival of wild species and especially, protected species, and the ability to respond to changing environmental conditions is expressed and maintained/conserved at genetic level. As noted above, the number of individuals may serve as a proxy for measuring genetic diversity.

Ecosystem means a dynamic complex of plant, animal, micro-organism communities and their interaction with abiotic features of the environment as a functional unit (see e.g. CBD, 1992). With a focus on individual organisms, habitat means the place or type of site where an organism or population naturally occurs (CBD, 1992). As jurisdictions may use habitat as another term for ecosystem (e.g. natural habitat types in European Union, 1992), we use the term habitat/ecosystem in this annex as a means of capturing all of these meanings. In the context of this annex ecological functions or habitat functions are those functions that an organism, population or community contributes to the habitat/ecosystem in which it resides.

Habitat structure comprises physical components of a habitat which are often formed by species and decomposing matter (e.g. standing or lying dead wood), but can also include abiotic features (e.g. gravel banks for spawning, water resources). Habitat structure is one of the key criterions for the assessment of habitat quality. Further criteria describing the quality of a habitat are habitat functions, typical species, the range and the area of the habitat.

Typical species are those frequently found in a habitat type or at least in a subtype or a variant of a habitat type (e.g. DG Environment, 2017).

Problem formulation

For this consideration, below are simple examples that illustrate the approach for planning an environmental risk/safety assessment. It includes a discussion of assessment endpoints, potential adverse effects, and a linear pathway to harm with corresponding risk hypotheses and information elements to illustrate the approach. As previously indicated in the document (section 1.2.6), the process is often more complex.

(a) Determination of assessment endpoints

A transgenic plant may have impacts on a protected species with which it interacts. At the species level, the attribute of the assessment endpoint is the number of individuals in (a) population(s) of the protected species under assessment, and not the ecological function the species fulfils, distinguishing this annex from the other environmental considerations discussed in this document (e.g. Annex C (Organisms (Animals)) or Annex D (Soil Functions)).

Maintaining the genetic diversity of a species or a population is an overall aim of conservation/protection of a species. Genetic diversity can be a specific, direct assessment endpoint for a species when it is possible to test the genotypes of individuals. This is more likely to be done in relation to conversation programmes. In a risk/safety assessment of a transgenic plant, as explained above, genetic diversity of a protected species can be indirectly assessed, for example, as (1) the abundance (the number of individuals in a given area), (2) the number and the size (number of individuals) of populations, and/or (3) the geographic distribution of the protected species. A transgenic plant may also influence a protected habitat/ecosystem, changing its species composition, structure, and/or quality. For example, if the transgene confers a characteristic that allows the transgenic plant to invade the protected habitat/ ecosystem, it may change the flora of the habitat/ecosystem and thereby, also the fauna of the habitat/ ecosystem, and finally the structure and/or quality of the habitat/ecosystem.

Two examples of assessment endpoints for the protected species and habitats/ecosystems as examples of the biodiversity environmental consideration are: (1) number of individuals in (a) population(s) of protected species under assessment, that as a proxy could correlate with genetic diversity of the protected species; and (2) species composition, structure and/or quality of the protected habitat/ecosystem.

(b) Identification of potential adverse effects on the assessment endpoints

The identification of potential adverse effects of a transgenic plant on biodiversity considers the characteristics of the transgenic plant linked to the genetic modification, including the novel gene product(s) or compounds and the potential receiving environments.

Any interaction of the transgenic plant with species or habitats/ecosystems dealt within the other annexes that has the potential to lead to adverse effects on protected species or habitat/ecosystems is, by nature, also relevant for biodiversity. The transgenic plant may alter the quality of a protected habitat/ecosystem located in the vicinity of a cultivation area by affecting the abundance or ecological functions of typical or vital species of a protected habitat/ecosystem, or by affecting the abiotic conditions of a protected habitat/ecosystem. As a result, the quality of the habitat/ecosystem may change.

Depending on the changed characteristics of the transgenic plant, in relation to the comparator that warrant further consideration, adverse effects on the assessment endpoints related to protected species and habitats/ecosystems may include: (1) reduced number of individuals in (a) population(s) of the protected animal species under assessment, that as a proxy could correlate with decreased/altered level of genetic diversity, of the protected animal species; and (2) changed species composition, structure and/or reduced quality of the protected habitat/ecosystem.

(c) Identification of plausible pathways to harm, formulation of risk hypotheses, and identification of information elements relevant to evaluating the risk hypotheses

In this section, two plausible pathways to harm are postulated. For each step of the postulated pathways to harm, a corresponding risk hypothesis is formulated that will enable the risk assessor to determine whether the pathways are likely to occur. Once it is shown that any part of a pathway is highly unlikely to occur, one does not need to continue evaluating the subsequent steps in the pathway and can conclude that the specific pathway to harm is unlikely to occur. In addition, examples of information elements that can be used to evaluate the risk hypotheses are given along with their rationales.

Because the adverse effects a transgenic plant can potentially have on a protected species, including their genetic diversity or habitat/ecosystem, may, in principle, happen via the same types of interactions as described in some of the other annexes, it follows that some of the same pathways to harm are also relevant. It should be noted that although the pathways and the risk hypotheses may overlap, the information elements may need to be adapted specifically to the protected species or habitat/ecosystem under consideration.

Postulated pathway leading to reduced numbers of individuals and genetic diversity of a protected animal species

Protected animal species occurring in the field or in field margins (i.e. managed ecosystems) and surroundings (e.g. unmanaged ecosystems) can interact with the transgenic plant through feeding. If the transgenic plant has a changed phenotype, this may lead to decreased survival, reproduction, fitness, and thus affect the abundance and genetic diversity of protected animal species. This is in principle the same as for any other species, as described in Annex C (Organisms (Animals)). However, in the case of a protected animal species, the attribute of the assessment endpoint is the number of individuals of the species itself and its genetic diversity and not the ecological function that the species in question provides. Hence, different information elements may be necessary.

One example of a postulated pathway to harm for the reduced numbers of individuals and genetic diversity of a protected animal species is shown in the first column of Table A G.1. Risk hypotheses for each step of the pathway are formulated in the second column and the third column provides examples of information elements for evaluating the hypotheses.

Table A G.1. Postulated pathway leading to reduced numbers of individuals and genetic diversity of a pollen-feeding protected lepidopteran species, corresponding risk hypotheses, and relevant information elements

Pathway steps	Risk hypotheses	Examples of information elements
In relation to the comparator, the transgenic plant produces a novel gene product	In relation to the comparator, the transgenic plant does not produce a novel gene product	Expression of a novel gene product in the transgenic plant
The novel gene product has a potential toxic effect on the protected lepidopteran species	The novel gene product has no potential toxic effect on the protected lepidopteran species	Sequence similarity to known toxic compounds Toxicity lab tests with surrogate species
The expression level of the novel gene product in the pollen of the transgenic plant adversely affects the protected lepidopteran species	The expression level of the novel gene product in the pollen of the transgenic plant does not adversely affect the protected lepidopteran species	Expression level of the novel gene product in pollen of the transgenic plant Comparison of expression level of the novel gene product in pollen and levels tested in toxicity lab tests with surrogate species
The larval food plant of the protected lepidopteran species is found in areas immediately adjacent to the transgenic plant	The larval food plant of the protected lepidopteran species is not found in areas immediately adjacent to the transgenic plant	Protected species location information and location of where the crop that contains the novel gene product will be grown
Pollen containing the novel gene product reaches the food plant of larvae of the protected lepidopteran species at levels that adversely affect the protected lepidopteran species	Pollen containing the novel gene product does not reach the food plant of larvae of the protected lepidopteran species at levels that adversely affect the protected lepidopteran species	Pollen dispersal characteristics of the transgenic plant Comparison of levels of the novel gene product expected to reach the larval food plant and levels tested in toxicity lab tests with surrogate species
There is overlap between the period of pollen shed from the transgenic plant and larval emergence of the protected lepidopteran species	There is not overlap between the period of pollen shed from the transgenic plant and larval emergence of the protected lepidopteran species	Knowledge of the phenology of the transgenic plant and of the protected lepidopteran species
An individual of the protected species ingests the novel gene product leading to decreased survival, reproduction, or fitness of the protected lepidopteran species	An individual of the protected species does not ingest the novel gene product and there is no decrease in the survival, reproduction, or fitness of the protected lepidopteran species	Number of individuals of the protected lepidopteran species
Number of individuals of the protected species decreases, and therefore genetic diversity decreases, due to decreased survival, reproduction, or fitness of the protected lepidopteran species		

It is important to note that examples of information elements in this table are intended to illustrate the types of information that can be used in evaluating a risk hypothesis, i.e. to determine whether particular pathway steps are likely to occur. However, for any step there might be other information that could be relevant. Rationales for how such information elements may be used to evaluate the risk hypotheses include:

- Information on where in the transgenic plant (e.g. pollen or leaves) the novel gene product is expressed and on the level of expression in the plant tissues, is useful in determining potential routes of exposure;
- Sequence similarity to known toxic compounds and toxicity lab tests with surrogate species provide information on potential adverse effects of the transgenic plant to the protected lepidopteran species;
- Toxicity testing of the novel gene product or of the pollen of the transgenic plant containing the novel gene product with surrogate species provide information on potential adverse effects of the transgenic plant to protected lepidopteran species;
- Protected lepidopteran species location information and location of where the crop that contains the novel gene product will be grown provide knowledge as to whether there is overlap between the habitat/ecosystem of the protected lepidopteran species and the area where the transgenic plant is grown;
- Pollen dispersal characteristics of the transgenic plant provides information on the potential for exposure;
- Comparison of levels of the novel gene product expected to reach the larval food plant and levels tested in toxicity lab tests with surrogate species provide information on whether potential adverse effects may result from the degree of exposure;
- Knowledge of the phenology of the transgenic plant and of the protected lepidopteran species provide information as to whether there is temporal overlap, e.g. timing of anthesis and/or larval emergence;
- A change in the number of individuals of the protected lepidopteran species can indicate decreased genetic diversity, survival, reproduction, or fitness of the protected lepidopteran species.

Postulated pathway leading to a changed species composition, structure, and reduced quality of a protected habitat/ecosystem

The cultivation of a transgenic plant may change the species composition, structure and reduce the quality of a neighboring protected habitat/ecosystem by affecting either species that are typical and vital for the specific habitat/ecosystem or by affecting functions and services that are vital for typical species in this protected habitat/ecosystem.

The postulated pathways leading to changes in the species composition, structure and reduced quality of protected habitats/ecosystems are in reality very complex. Many pathways overlap, pathways may have several branches and some information is not easily obtained.

One example of a postulated pathway to harm for changed species composition, structure, and reduced quality of a protected habitat/ecosystem is shown in the first column of Table A G.2. Risk hypotheses for each step of the pathway are formulated in the second column and the third column provides examples of information elements for evaluating the hypotheses.

Table A G.2. Postulated pathway leading to changed species composition, structure, and reduced quality of a protected habitat/ecosystem, corresponding risk hypotheses, and relevant information elements

Pathway steps	Risk hypotheses	Examples of information elements
The transgenic plant is a tree, and the introduced trait confers increased tolerance to drought stress due to water use relative to the comparator	The introduced trait does not confer increased tolerance to drought stress due to water use relative to the comparator	Function of the introduced gene and associated phenotype of the transgenic tree
The transgenic tree is planted in the area of a protected habitat/ecosystem	The transgenic tree is not planted in the area of a protected habitat/ecosystem	Knowledge of locations of protected habitats/ecosystems; Visual observation of protected habitats/ecosystems
Seeds of the transgenic tree are spread in a neighbouring protected habitat/ecosystem.	Seeds of the transgenic tree are not spread in a neighbouring protected habitat/ecosystem	Existence of protected habitats/ecosystems in the area of cultivation; Seed dispersal distance
The transgenic tree has a fitness advantage in a protected habitat/ecosystem compared to the comparator due to water use (e.g. roots extend deeper into the soil and hence access water table better than comparator).	The transgenic tree does not have a fitness advantage in a protected habitat/ecosystem compared to the comparator due to water use	Presence of drought stress in a protected habitat/ecosystem Vegetative and reproductive performance of the transgenic tree and the comparator in the presence of drought stress
The number of self-sustaining populations of the transgenic tree increases in the protected habitat/ecosystem compared to the comparator	The number of self-sustaining populations of the transgenic tree does not increase in the protected habitat/ecosystem compared to the comparator	Establishment and persistence of populations of the transgenic tree in habitat/ecosystem types similar to the protected habitat/ecosystem compared to the comparator
The quality of the protected habitat/ecosystem is reduced through reduction of the water table	The quality of the protected habitat/ecosystem is not reduced through reduction of the water table	Levels to which the water table in areas of cultivation of the transgenic tree is reduced
The abundance of typical plant species of the protected habitat/ecosystem decreases due to the decrease in the water table	The abundance of typical plant species of the protected habitat/ecosystem does not decrease due to the decrease in the water table	Typical plant species of the protected habitat/ecosystem and their ecology particularly with respect to their water consumption needs
Species composition and structure are changed, and quality reduced in the protected habitat/ecosystem		

It is important to note that examples of information elements in this table are intended to illustrate the types of information that can be used in evaluating a risk hypothesis, i.e. to determine whether particular pathway steps are likely to occur. However, for any step there might be other information that could be relevant. Rationales for how such information elements may be used to evaluate the risk hypotheses include:

- Function of the introduced gene and associated phenotype of the transgenic tree provide information on the potential for displaying increased resistance to drought stress due to water use relative to the comparator;

- Existence of protected habitats/ecosystems in the area of cultivation and seed dispersal distance provide information on whether seeds of the transgenic tree may enter in the protected habitats/ecosystems;
- Presence of drought stress in the protected habitat/ecosystem and vegetative and reproductive performance of the transgenic tree and the comparator in the presence of drought stress provide indication of the potential for increased survival and reproduction in the presence of the abiotic stressor;
- Establishment and persistence of populations of the transgenic tree in a habitat/ecosystem types similar to the protected habitat/ecosystem provide information on whether the potential for development of self-sustaining populations of the transgenic tree is increased in the protected habitat/ecosystem;
- Levels to which the water table in areas of cultivation of the transgenic tree is reduced;
- Typical plant species of the protected habitat/ecosystem and their ecology particularly with respect to their water consumption needs;
- Species composition and structure development in comparable habitats/ecosystems if tree populations establish there provide information on the potential that this may happen in the protected habitat/ecosystem as water table falls.

References

CBD (1992), *Convention on Biological Diversity*, United Nations, https://www.cbd.int/doc/legal/cbd-en.pdf (accessed 26 April 2023).

Commonwealth of Australia (1999), *Environment Protection & Biodiversity Conservation Act 1999*, The Federal Register of Legislation at https://www.legislation.gov.au/Details/C2022C00214 (accessed 26 April 2023).

DG Environment (2017), Reporting under Article 17 of the Habitats Directive: Explanatory Notes and Guidelines for the period 2013–2018, Final version – May 2017, European Environment Agency, Brussels, Habitats Directive reporting resources (europa.eu) (accessed 26 April 2023).

European Union (2018), Commission Directive (EU) 2018/350 of 8 March 2018 amending Directive 2001/18/EC of the European Parliament and of the Council as regards the environmental risk assessment of genetically modified organisms, https://eur-lex.europa.eu/legal-content/EN/TXT/?uri=celex%3A32018L0350 (accessed 7 July 2023).

European Union (1992), Commission Directive 92/43/EEC of 21 May 1992 on the conservation of natural habitats and of wild fauna and flora, https://eur-lex.europa.eu/eli/dir/1992/43/2013-07-01 (accessed 26 April 2023).

United States Department of the Interior (1973), Endangered Species Act of 1973 as amended through the 108th Congress, U.S. Fish and Wildlife Service, Washington D.C., https://www.fws.gov/media/endangered-species-act (accessed 26 April 2023).

Annex H. List of OECD consensus documents on environmental safety assessment, 1996-2023

Consensus document	Lead country(ies)	Year of issue	Volume
Facilitating harmonisation			
Designation of a Unique Identifier for Transgenic Plants – 2006 revised version (guidance document)	Working Party	2006	Vol. 3
Introduction to the OECD Biosafety Consensus Documents – *updated for each volume*	Working Party	2005	Vol. 1, 3, 4, 5, 6, 7, 8, 9
Low-Level Presence of Transgenic Plants in Seed and Grain Commodities: Environmental Risk/Safety Assessment, and Availability and Use of Information	Working Party	2013	Vol. 6
Molecular Characterisation of Plants Derived from Modern Biotechnology	Canada	2010	Vol. 3
Revised Points to Consider for Consensus Documents on the Biology of Cultivated Plants – *replacing the 'Points to Consider' section of Vol.3*	Working Party	2020	Vol. 9
Environmental Considerations for Risk/safety Assessment for the Release of Transgenic Plants	Working Party	2023	Vol. 10
Traits			
Crop Plants Made Virus Resistant through Coat Protein Gene-Mediated Protection	Task Group	1996	Vol. 1
Genes and their Enzymes that Confer Tolerance to Glyphosate Herbicide	Germany, Netherlands, United States	1999	Vol. 1
Genes and their Enzymes that Confer Tolerance to Phosphinothricin Herbicide	Germany, Netherlands, United States	1999	Vol. 1
Herbicide Metabolism and the Residues in Glufosinate Ammonium (Phosphinothricin) – Tolerant Transgenic Plants	Germany	2002	Vol. 1
Transgenic Plants Expressing Bacillus thuringiensis Derived Insect Control Protein	United States	2007	Vol. 3
Micro-organisms			
Information used in the assessment of environmental applications of micro-organisms			
Acidithiobacillus	Canada	2006	Vol. 2
Acinetobacter	Canada	2008	Vol. 4
Baculovirus	Germany	2002	Vol. 2
Pseudomonas	United Kingdom	1997	Vol. 2
Guidance documents on biosafety aspects of bacteria			
Horizontal Gene Transfer Between Bacteria	Germany	2010	Vol. 4
Methods for Detection of Micro-organisms Introduced into the Environment: Bacteria	Netherlands	2004	Vol. 4
Use of Information on Pathogenicity Factors: Bacteria	Canada, Netherlands	2011	Vol. 5
Use of Taxonomy in Risk Assessment of Micro-organisms: Bacteria	Canada, United States	2003	Vol. 4

Biology of crops			
Apple (*Malus domestica*)	Belgium, Germany	2019	Vol. 9
Bananas and plantains (*Musa* spp.)	Spain	2009	Vol. 4
Brassica crops (*Brassica* spp.) – replacing, and completing with other species, the Oilseed rape chapter of Vol.1	Canada	2012	Vol. 5
Cassava (*Manihot esculenta*)	Brazil, AUDA-NEPAD, ILSI-CERA	2014	Vol. 6
Chili, hot and sweet peppers (*Capsicum annuum*)	Korea, Mexico, United States	2006	Vol. 1
Common bean (*Phaseolus vulgaris*)	Brazil, ILSI-CERA	2015	Vol. 6
Cotton (*Gossypium* spp.)	Spain	2008	Vol. 4
Cowpea (*Vigna unguiculata*)	Australia	2015	Vol. 6
Maize (*Zea mays* subs. *mays*)	Mexico	2003	Vol. 1
Oyster mushroom (*Pleurotus* spp.)	Korea	2005	Vol. 1
Papaya (*Carica papaya*)	United States	2005	Vol. 1
Potato (*Solanum tuberosum* subsp. *tuberosum*)	Netherlands, United Kingdom	1997	Vol. 1
Revised Rice (*Oryza sativa*) – replacing the Rice chapter of Vol.1	Japan	2021	Vol. 9
Safflower (*Carthamus tinctorius*)	Australia	2020	Vol. 9
Sugar beet (*Beta vulgaris*)	Switzerland	2001	Vol. 1
Sugarcane (*Saccharum* spp.)	Australia	2013	Vol. 6
Sunflower (*Helianthus annus*)	France	2004	Vol. 1
Sorghum (*Sorghum bicolor*)	South Africa, United States	2016	Vol. 7
Soybean (*Glycine max*)	Canada	2000	Vol. 1
Squashes, pumpkins, zucchinis and gourds (*Cucurbita*)	Mexico, United States	2012	Vol. 5
Tomato (*Solanum lycopersicum*)	Mexico, Spain	2016	Vol. 7
Wheat (*Triticum aestivum*)	Germany	1999	Vol. 1
Biology of trees			
Timber trees			
Birch: European white birch (*Betula pendula*)	Finland	2003	Vol. 2
Douglas fir (*Pseudotsuga menziesii*)	Canada	2008	Vol. 3
Eucalyptus (*Eucalyptus* spp.)	Australia	2014	Vol. 6
Larches: North American larches (*Larix lyalli, Larix occidentalis, Larix laricina*)	Canada	2007	Vol. 3
Pines: Eastern white pine (Pinus strobus)	Canada	2002	Vol. 2
Pines: Jack pine (*Pinus banksiana*)	Canada	2006	Vol. 3
Pines: Lodgepole pine (*Pinus contorta*)	Canada	2008	Vol. 3
Pines: White pine (*Pinus monticola*)	Canada	2008	Vol. 3
Poplars (*Populus* spp.)	Canada	2000	Vol. 2
Spruces: Black spruce (*Picea mariana*)	Canada	2010	Vol. 3
Spruces: Norway spruce (*Picea abies*)	Norway	1999	Vol. 2
Spruces: Sitka spruce (*Picea sitchensis*)	Canada	2002	Vol. 2
Spruces: White spruce (*Picea glauca*)	Canada	1999	Vol. 2
Fruit trees			
Apple (*Malus domestica*) [also listed above in "Biology of crops"]	Belgium, Germany	2019	Vol. 9
Bananas and plantains (*Musa* spp.) [also listed above in "Biology of crops"]	Spain	2009	Vol. 4
Papaya (*Carica papaya*) [also listed above in "Biology of crops"]	United States	2005	Vol. 1
Stone fruits (*Prunus* spp.)	Austria	2002	Vol. 2
Biology of animals			
Atlantic salmon (*Salmo salar*)	Finland, Norway, United States	2017	Vol. 7
Mosquito *Aedes aegypti*	Brazil, Mexico, ILSI-RF	2018	Vol. 8

www.ingramcontent.com/pod-product-compliance
Lightning Source LLC
LaVergne TN
LVHW060455061126
851857LV00016B/1842
* 9 7 8 9 2 6 4 8 9 7 9 6 0 *